AF765183

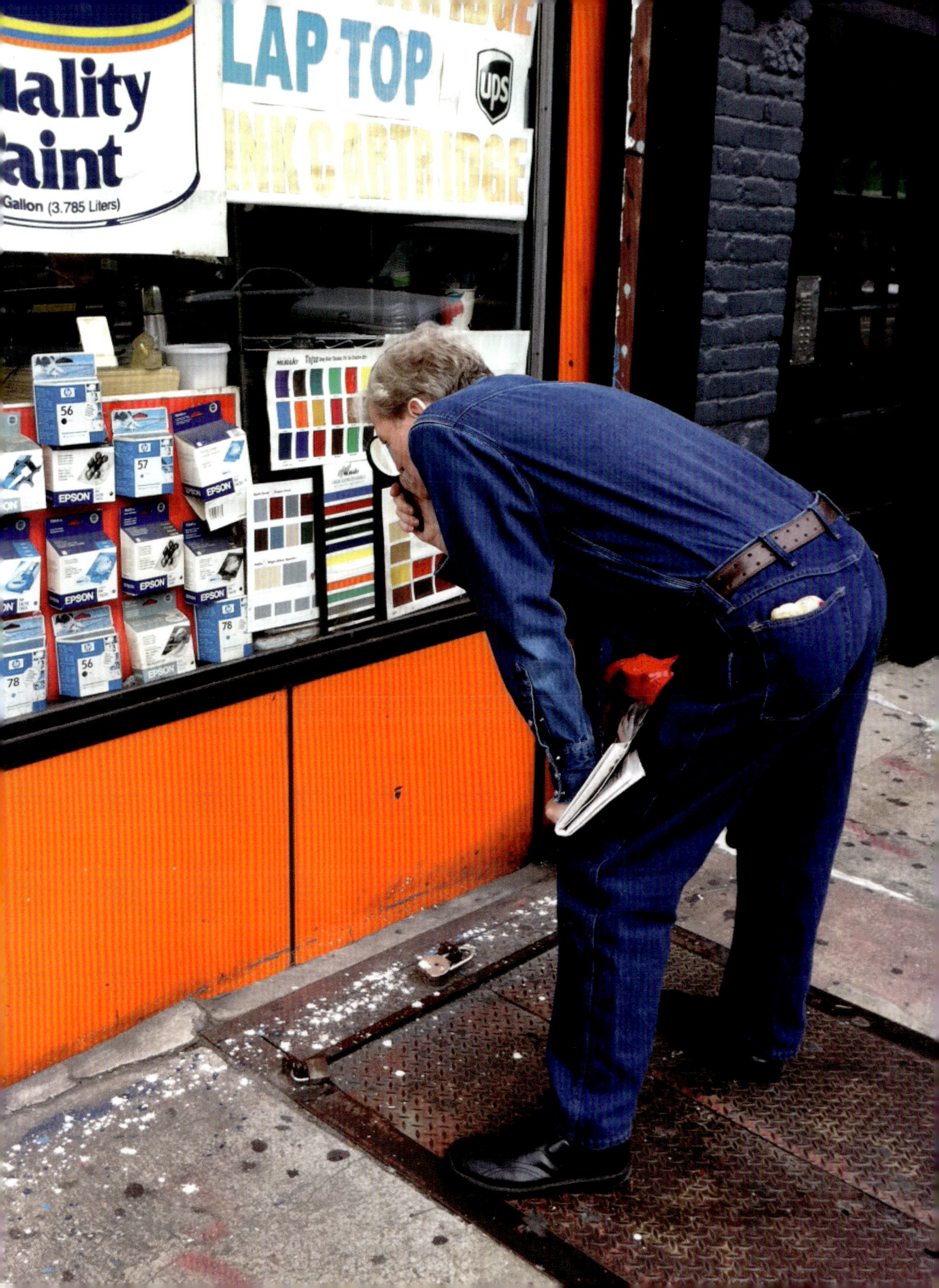
uality
aint
Gallon (3.785 Liters)
LAP TOP
ups
INK CARTRIDGE
EPSON
56
57
78

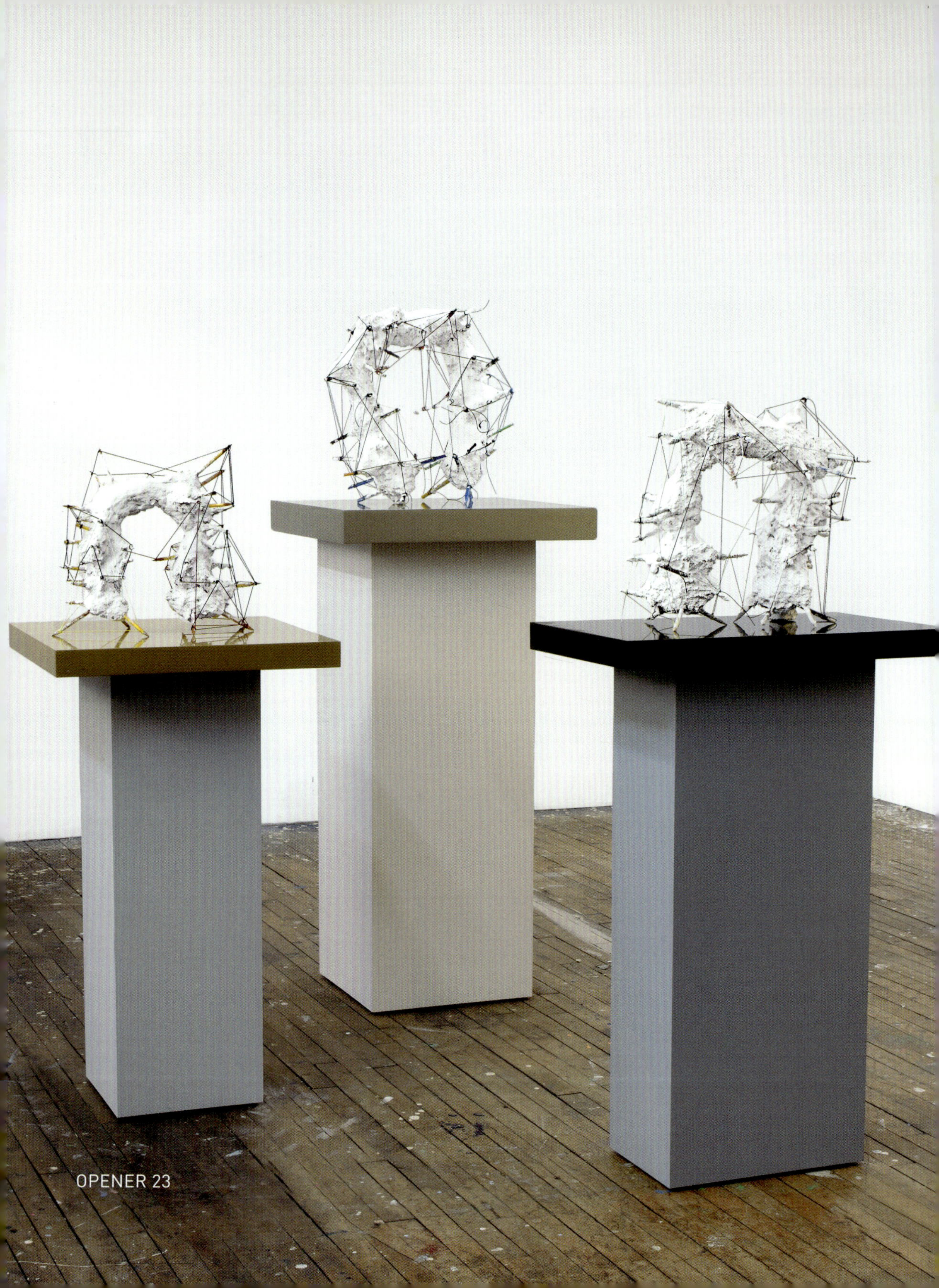

PAM LINS
DENVER GOLD

IAN BERRY

THE FRANCES YOUNG TANG
TEACHING MUSEUM AND ART GALLERY
SKIDMORE COLLEGE

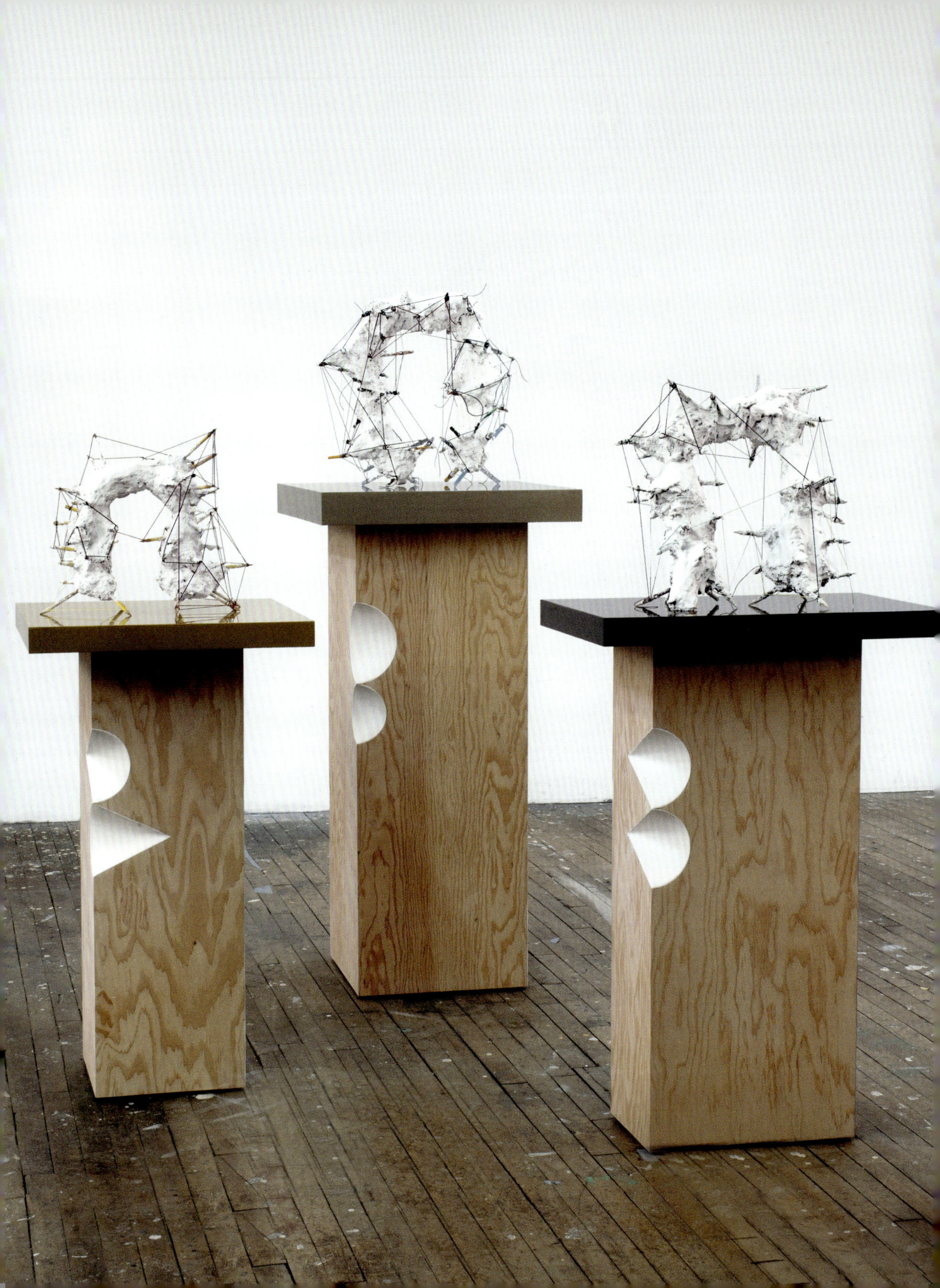

A BLURRED POSITION

A DIALOGUE WITH PAM LINS BY IAN BERRY

PAM LINS REFERS TO HER WORK primarily as sculpture and consistently interrogates the medium through its relationship to other forms of representation, most notably painting and photography. Mixing a love of formal, modernist art with an idiosyncratic sensibility, Lins's work complicates divisions between abstract and representational, colorful and unadorned, hollow and full, and front and back.

Constructed in response to the Tang Museum's mezzanine space, *Denver Gold* continues Lins's long-running exploration of sculpture's unique relationship with its viewers and its environment. At the center of the installation lies a forest of freestanding sculptures whose plywood pedestals, topped with small paintings and objects, play with traditional conventions of sculptural display. In the re-created *Corner piece reconstructed painting background*, a section of the wall remains unpainted, revealing its internal material. On a curved shelf, paintings, propped as though still in the studio, gradually recede outward until the final pieces hang on the finished gallery wall.

Several works refer to photographs of sculpture by well-known modernist sculptors, including Constantin Brancusi (1876–1957), Naum Gabo (1890–1977), and Henry Moore (1898–1986). Lins's dialogue with these influential figures of modern sculpture comes from a deep admiration of their work, as well as a keen understanding of her relationship to it as a contemporary female sculptor. For *Fireplace reconstructed sculpture background,* Lins recreates a section of Brancusi's studio. An avid photographer of his own work, Brancusi used a monochromatic painting propped behind his fireplace as a backdrop

Opposite and previous:
Studio views, 2012

for photographing his sculptures. Lins's newest series, *Slabs and Armatures*, similarly shine light on the unseen, eccentric aspects of art-making—the plaster sculptures refer to a photograph Lins found of an armature inside a Moore sculpture. These pieces sit upon a shelf-like base painted in exterior color options for the 1974 Pontiac GTO. Their shades and names conjure an earlier time in American visual culture—for Lins, specifically, her childhood in the Midwest. In this way Lins reclaims the less-examined parts—the inside, the middle—of artistic influence and experience.

Ian Berry: Many of your works explore how we look at a sculpture, taking on some of the big ideas in the history of sculpture while also inventing ways not to be beholden to that history.

Pam Lins: I know it seems limiting to stick with the genre of "sculpture" at this point, but I see my project as more reformatory in approach. I consider other forms and languages that frame and depend on the category of sculpture but also on the conversations floating around in real time.

IB Many of your works explore notions of display.

PL **Yes, and of course this seems a bit obvious with sculpture. We're tied at the hip. I remember at one point I was showing photographs of my work and someone said, "Maybe next time you shoot it, you should take it off the kitchen table." How do you respond to this? That table was part of my sculpture. Photography has always been the primary medium sculpture has wrangled with. I'm interested in display, in the whole predicament of getting the thing off the floor or onto the wall or whatever, and the language of display in relation to sculpture but also as a discussion of how we live. I think of these locations and situations as a tool to discuss space within and between sculpture and the built world.**

IB For you the display isn't separate?

PL **No. It's a smudge, an indistinct mass.**

IB You're just making a thing.

PL **Yes, but it took me a while to get there.**

IB With your earlier work you would make a sculpture, and make a display after?

PL **It was forced, but it was always something I was interested in. I think that I'm there now.**

IB What makes you feel like you're there—what is different?

PL **For me it's because I've been making work for a good chunk of time. Being able to look back and forward helps the way I work. I think it is also the product of trying things, and the idea that if you say something enough you start to believe it yourself. When I hear someone describe my work as *atop* something or *beside* something, or using these kinds of adjectives of display, I always get a little upset because my working mode deals with this idea of blurring until work and display become one. The pedestal becomes the sculpture, the thing on top, the thing below, the stuff from the inside. I like the idea of becoming. The painting on the wall, which is now in the sculpture, speaks to modes of perception. So I like to think about it as a kind of blurred position.**

IB I love the idea of a blurred position. You have made a number of works in the past few years based on looking at older sculptors' work or classical sculpture.

PL **Old men.**

IB How do you pick them?

PL **I have spent a lot of time with some of them, like Brancusi. I've been interested in him for a million reasons. Then there are people like Henry Moore, whose work I don't know if I always like, but who is just as interesting to me to use, and claim, to rethink some problems. I like to hunt through archives and collections of pictures of artists in their** studios and try to find some weird thing that's like the threshold of their work—something that has never been historically organized—and work with that. It involves looking at a ton of photographs, so sometimes it's about who you can find photos of. All the old men have been archived. So, in a way, I'm also speaking to who has been left out.

IB Is it less about their artwork and more about the archive - those pictures you find?

PL I can't say that across the board, but it is about how they worked in their studios and about the photo archives, and how to think about what a photo is.

IB What's so great about time in the studio?

PL Nothing; it's often horrible! It's hard. And I respect it as a way to try to get off the computer, bring my body back from the screen freeze. I know that when I go into my studio I don't have that much time because of the crazy way we live. I'm super-interested in how work is generated, often more than the final product. I'm often disappointed in the final product.

IB You like the process?

PL Yes, totally. I don't go into the studio to close out the world and have a solitary experience. I go into the studio to let the world in and contemplate everything. It's just quieter; it's a quiet place for me. Except talk radio is always on, and from that the news seeps in. And also music. I listened to the countdown of five hundred classic rock songs over Memorial Day and was surprised to realize I knew five hundred songs. New thoughts meet old thoughts.

IB What does it mean to let the world in?

PL I don't go in there to be alone and go deep into my unconscious although I may hope and wait for that to happen. It's not about getting away. It's not about being alone. Alone, but not lonely. It's not about some kind of...

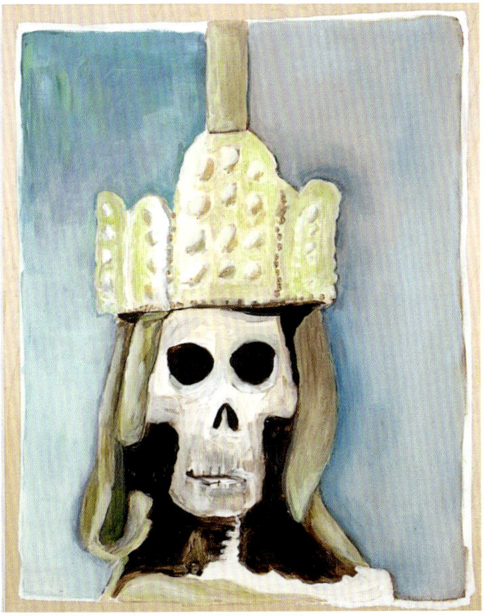

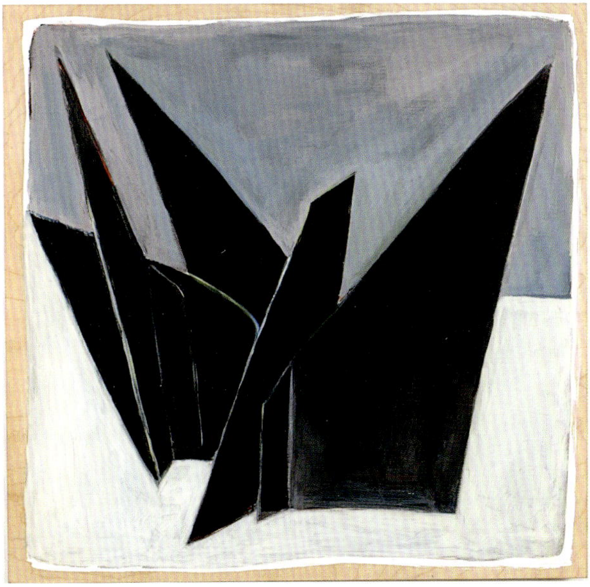

(left)

Charles the VI layout and fog, 2011
Oil on panel
18 x 14 inches

(right)

Gwen Fritz Delay, 2011
Oil on panel
16 x 16 inches

IB Romantic isolation?

PL **Right. It's really to sort out the world—to edit, read, generate thoughts, pick through things. Proustian and Freudian.**

IB You have an activist side. Do you see your work dealing with older male artists as a part of that? How have you made your activism fit with your artwork?

PL **All histories should be questioned. Don't you have to put yourself in a sort of contradictory position of studying and questioning simultaneously? And to forget both at times? I depend on my gaps to make work. I know this sounds antithetical to how we are supposed to proceed, but I want to use history like information or material... sometimes! I've always been blown away that the study of art history and the practice of being an artist are placed next to each other sometimes, and separated at other times. I also think aesthetics and politics can lie together.**

IB But you don't make work that's narrative or has a story?

PL **No.**

IB What fuels your work?

PL **It's fueled by this idea of questioning history. This is why I often remake work, which I think at some level is an empathetic process: to know what it feels like to remake someone else's work, or be near that feeling. The remaking I'm involved in doesn't mean a copy. I like to have** that knowledge of what it would feel like to make someone's work and how that can generate an impression that influences the way I look at that work, think about the world, make my own work, remind myself not to be an asshole, think about what's next, and project my work out there. If I make something and you look at it—something has happened.

IB You've also remade your own work.

PL Yes. For me, it's a way to be a viewer of my own work. A device to split the self.

IB It's an alternative way of working because most of us value studio time as time to make something new. Deciding to remake something is the opposite—it's going in the opposite direction of what a lot of the world is telling you. In that way, your practice involves a bit of resistance.

PL It is a bit of resistance but probably like aggressive resistance and some sort of re-knowing of resistance. I don't know what it means to remake a work but I think it's good. There are always changes in the remaking. I'm not an artist that's constantly in touch with the meaning of the work.

IB Can you describe how remaking operates in an earlier work, like *Please Bear With Us*?

PL In the works *Please Bear With Us,* I employed a sort of remaking. I could barely find any women artists while using the archives, so I remade some work by some contemporary women painters and also sculptural works based on mythic figures: Mary Magdalene and Medusa. I think about the stasis of history versus the action of the recent past. One of these pieces is an homage to Alma Thomas and the other is for Moira Dryer. It comes back to the idea of remaking and what homage is. I know homage is a word that people don't really use anymore, which of course is partly why I want to drag it out. It satellites around other words; influence, appropriation, quotation, grabs, loving, stealing. I think about remaking as an event. Maybe it could offer something emotional.

IB Putting homage back in circulation?

PL Yes, I would like to see someone write about homage right now. I think I approached the women painters in a similar way as I was doing those other men: if I remake their thing, I have a simple empathy and a re-knowing and an understanding. There is a circular psychology about this process for me—I can see the painting; can it see me if I make it? It is a condition of maintaining an attachment to a problematic object in advance and while you are remaking it.

IB Can you say more about an artwork seeing you?

PL There are a couple of things by Lacan that have stuck with me. One of the reasons I remember Lacan is because among his weird, metaphysical style of writing and my "creative memory" there are some really great anecdotes. In one of them, he is on this fishing boat and the guy he is with points to a shiny tuna can floating in the sea and says, "You see the can? Do you see it? Well, it doesn't see you!" I just love that. The process of remaking is a little bit like this for me—I can see their work, but until I think about a remake or find the right item, I'm not really part of it. It's hard to find the right situation. But it also makes me think of cans all the time.

In another anecdote, Lacan is describing a pigeon that won't mature into an adult until it recognizes its own species. Then they figured out that if they held a mirror in front of the pigeon it would mature, so it just needed to see another version of itself. He suggests that the mind needs this image to organize the visible world—the subject must be able to visualize herself in relation to the objects around her. That's my project! and I'm not a Lacanian. It's an enlightening moment in a portion of a text. Sometimes those anecdotes get close to how I feel. I think it's some kind of mix of splitting yourself, trying to remake yourself, something about how others feel, and something about the politics of feminism, and some way to consider abstract political thought. But I also love and use comedy and humor and Buster Keaton in the same way. This is probably

more improv, as I think scripted comedy may have more to do with craft. My work is very concerned with how things are made, but I am more interested in trying to keep the work moving somehow.

IB How do you see feminism working in that mix?

PL **Feminism has such a complicated history that it's hard to figure out how to be creative with it. That's the key: how to take these things with these impassioned histories that you don't want all of, and figure out how to work with them creatively at the same moment you are studying them. I allow myself some abstract thinking.**

IB That's a great way to use something in a building way. There are lots of ways that transaction can get bogged down by arguing about who is right and who is wrong, and not really moving forward.

PL **Artists always try to stick stuff out into the world that people aren't used to.**

IB You've taught for many years—do you have a clear break between your teaching life and your art life?

PL **No. I don't teach as an extension of my practice by any means, but it puts me back in touch with the reasons I wanted to become an artist.**

IB What are those reasons?

PL **I feel that art can help the human condition. I try to hold onto those ideals even though we have to negotiate all the problems with the market and all that distracting stuff. I still feel it's something that can help us, though I shouldn't use the word help. It teaches us that we're asking the wrong questions.**

IB At the Tang, we considered how to position the sculptures and how people would move through the space. What do you think about when you're setting up an installation? Your last few exhibitions have felt like total artworks in themselves.

PL I think every artist would probably rather work in a space, in situ. The space of the gallery I show at, Rachel Uffner, is a Lower East Side storefront, so it's long and skinny. Because I have been in New York for a long time, it's like a historical space. Before her gallery was there it was the artist-run gallery Orchard, which I had been to quite a bit, and before that it was a store. So dealing with a space that I have been to in other incarnations is weird in itself. It's a container. I think that also loads the space for me. It's the fascinating and sometimes horrible thing about New York: how it changes. I'm sure you feel that way with the Tang. Sergej Jensen is an artist who hangs his show the way the show before was hung. That seems okay to me; in fact, I think it is also hilarious. But for my installations, or the installing of the work, I mostly consider the walk.

One of the things that I take on with sculpture is the idea that you walk around it. And this is also a clichéd negotiation about looking at sculpture. But the space around a sculpture is much more unruly than the space of a sculpture. I try to consider it an optimistic walk because you are going into the future and you don't really know what's around the corner. There is an idea of hopefulness. And once you pass that corner, a memory kicks in, which I think about in terms of both the body and the face, each of which brings up a different sort of response. Ideas about beauty are shaped in part by notions of scale and symmetry derived from the dimensions of the human body. With the works that picture a kind of portrait head, such as Medusa [*Medusa page design decoy*, 2010] or Lincoln [*Lincoln bookend obstruction*, 2010], this experience is delayed a bit when the beauty of a face is suddenly presented. The sculptures' placement obscures things, delays things, and blocks things, all as you're walking. I work with that.

IB Like a speed bump, it slows you down.

PL It could be like a speed bump, but I don't think about it as manipulating. It's more related to seeing and considering suspension as you move. It's more like walking through a Japanese garden but not taking the

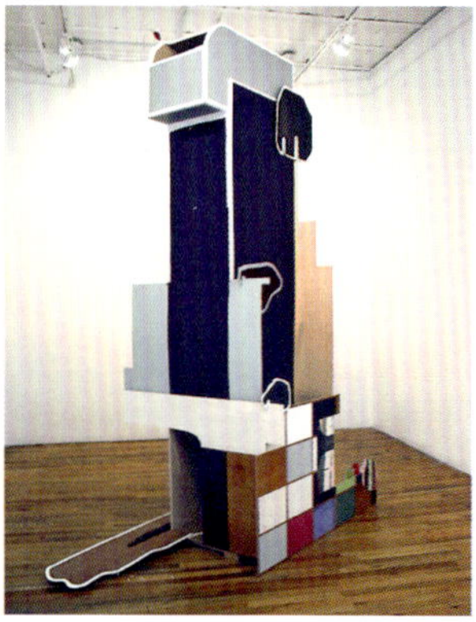
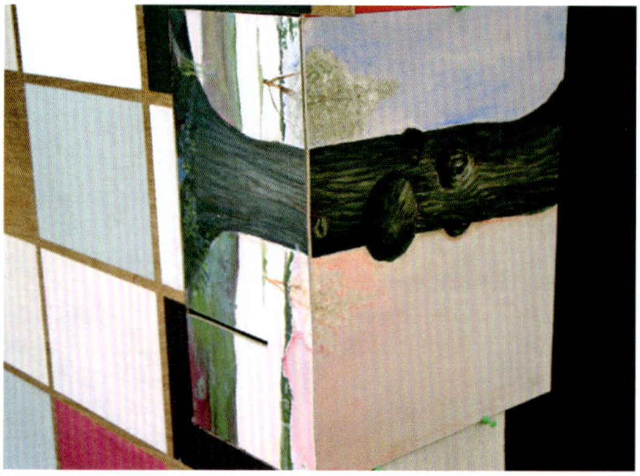

Everything I know about the tree I learned from the mailbox, 2005 (overall and detail)
Wood, cardboard, acrylic, paper

path—when you walk here, you'll see this, and it will be framed by that, and then you'll walk over here and get this view based on the memory of the last view. It's not necessarily about slowing people down, although I try to throw in delays.

IB That's more of an architectural point of view.

PL **It's about the body walking through built space, so in that way it does have to do with architecture. Sculpture always considers the spaces around it. I think it's a sort of a compressed dialogue between the body, the sculpture, and the space.**

IB There are many cutouts in your work, like the white recesses on the sides of your wooden sculptures.

PL **The niches.**

IB They are holes?

PL **Yes, there is some idea of a hole, which is something that I have a bit of a relationship with. There is an index in my work for dealing with the interior of a sculpture, and interiority. There was a time when not all sculpture was hollow. What does it mean to have hollow sculptures? I** always like to touch on that. I went from being interested in infrastructure within buildings (like ventilation) to more provisional situations, like slitting and slotting and rigging. Now I am sort of digging out and building back up. It's about negative space. It's about the politics of being hollow. Eventually hollow processes of sculpture moved weight and material without showing the loss. What to do with the loss....

IB Are you taking that negative space and reinserting it somewhere else?

PL There is some of that. It's never that easy because I look for those spaces in the world and sometimes those spaces are a negative of a sculpture and sometimes it's a hole. Considering and revealing the interiority of a base engages the process of its exposing itself—redirecting the interiority. It is showing itself while remaking and extending itself. Then there is active hollowness. The digging of the New York City subway tunnel moved the waste from the hole around. The excavation removed internal debris and rock and the foundations from skyscrapers and moved it horizontally or above ground to extend Governors Island and the Jersey shore. The comings and goings of all of the conditions and networks of the subway revolve around that hole.

IB Let's talk about your paintings. You paint from photographs, some that you take and some that you find?

PL Painting has a lot of different sources for me. I often get into a project by painting my way into it. As I'm researching all those crazy archives and picking out pictures of artists in their studios, I'm painting pictures of their work. I create a transcription. Some of that makes it to the final work, some of it gets painted over, and some of the representation gets abstraction over it. I think in some ways it also has to do with why I started painting. It's kind of corny, but this idea that with paint and a couple movements of my hand, I can do a whole lot—that means a lot in contrast with making sculpture, which is heavy, slow, and a total pain. It has to do with time. You can have a quicker thought that you get out. And it can change the time and space of the read of a sculpture.

IB Is it like notetaking?

PL It can be related to that. Or jotting. Accumulative jottings can add up to a lot after awhile. That may be my work method. Then I stuck the paintings in the sculptures. Painting has also always been my alpha, and sculpture was often defined by how it worked in regard to painting (many times negatively). I always felt just in awe standing in front of paintings. Many of the really smart people I know are painters, and painting's history is so long and articulated and written about on so many levels. I just always went to painting. Maybe every artist has to. Finally I just made the move and started painting in 2004. I was a horrible painter, although there is nothing I like better than a bad painting. I have looked at thrift store paintings for a long time.

IB Why do you love them?

PL I don't know. Maybe because I grew up in the Midwest and that was what was around. I loved Jim Shaw's *Thrift Store Paintings* and used to comb through stores in Chicago, looking for things. I began painting from my photo archive. I'm a better painter now and am starting to deal with painting in ways similar to my interests in sculpture. I'm trying to think about the future condition and about time, because a painting could almost be a stand-in for an armature in a sculpture, suggesting a future. A few years ago I started calling the paintings in my work armatures, I had been looking at an archive of photos of Avard T. Fairbanks working in his studio on sculptures of Lincoln. There are many photos of the armatures before he applied the body. I made paintings from some of the photos of these armatures. Things hang on them. They support things. They are a foundation for something else beside themselves. They hold a language. The representation of an armature became a schema that links a shifting infrastructure to an idea of a real thing. I'm not thinking just about judging sculpture using pictorial criteria, but more to disrupt and delay and blur the way I both use and avoid an essentialist read of sculpture. The transcripted paintings and the sculptures they are part of can exist in a degraded space. Then I realized it didn't need to be a painted representation of an armature. Just a painting in a sculpture could grip that logic. So they could be relieved a little bit....

IB Of painting's history?

PL **Yes, and of their finished-ness.**

IB Can you describe what gets lost when something is finished?

PL **That's a really good question. I think that's the answer. Something gets lost when you finish it. The question is the answer.**

IB There is a melancholy about having to leave something.

PL **Right, and you're confronted with this idea that you have to move on.**

IB You have to finish, and finishing doesn't feel right.

PL **It is never good when you finish the meal, or when you finish the movie, or finish the book. I'm not even going to go to the sex thing.**

IB A lot of artists, curators, and writers try to suggest that their work is open as a way to avoid this problem.

PL **I think it's related to that. You did just call it a problem. It's also about the fact that you can wake up in the morning and think differently about the way you felt about something the day before.**

IB You want to stay alive.

PL **That's for sure. You want to stay alive. You want to be in the present.**

IB Do you think about viewers when you make something?

PL **Yes, but I've also always done things my own way and been very confused by that. I think that probably brings us back to why I remake work, because I want to know what it feels like. I've often thought that, and this is something I talk about when I teach, that it's one thing to make work, and you can probably sort of figure that out, like figure**

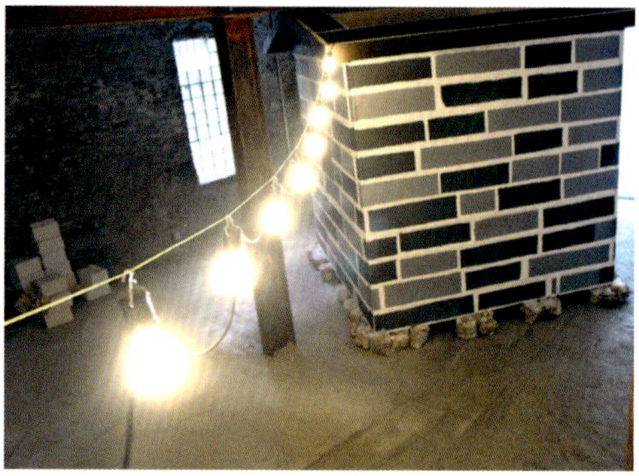

out how to be an artist. But I think being a really active and aggressive viewer and spectator is super hard. How do you become the unconstrained viewer? It's a lifelong thing that I struggle with. I don't know how to be empathetic with a viewer that well. But that's part of what keeps me going.

IB You have close relationships with other artists. Are they in your head when you're in your studio making things?

PL **I hope that I'm generous to the people that have looked at a lot of my work, that I've had a lot of conversations with, that I've gone to their studios, that I've been influenced by their ideas. I really hope that I'm generous to them, or at the very least that they think I'm funny.**

IB I think a community of artists is very important. To many artists the community of artists that they are in, whether those artists are alive or dead, or people they know or don't know, or their partners or their studio mates; that community is critical. I think that's often invisible to viewers.

PL **Hearing you say that reminded me of what someone said the other day: that he wanted to make his work for the average Joe, and I was thinking to myself, well I'm the average Joe, too. I never want to have that kind of bigotry with regard to who does and doesn't understand my work.**

IB If you didn't have viewers, would you still make art?

PL **You want to show your work. You want to put eyes on the work. It goes back to what we talked about earlier—how I think art can live in the world and help things. I am an artist who has made a lot of work that was never shown. I had to think very creatively about storage. Barely anyone sticks with sculpture.**

IB It's public speaking.

PL **I hope so. You strive to know more. It keeps you there. I keep learning things and doing things, and giving things.**

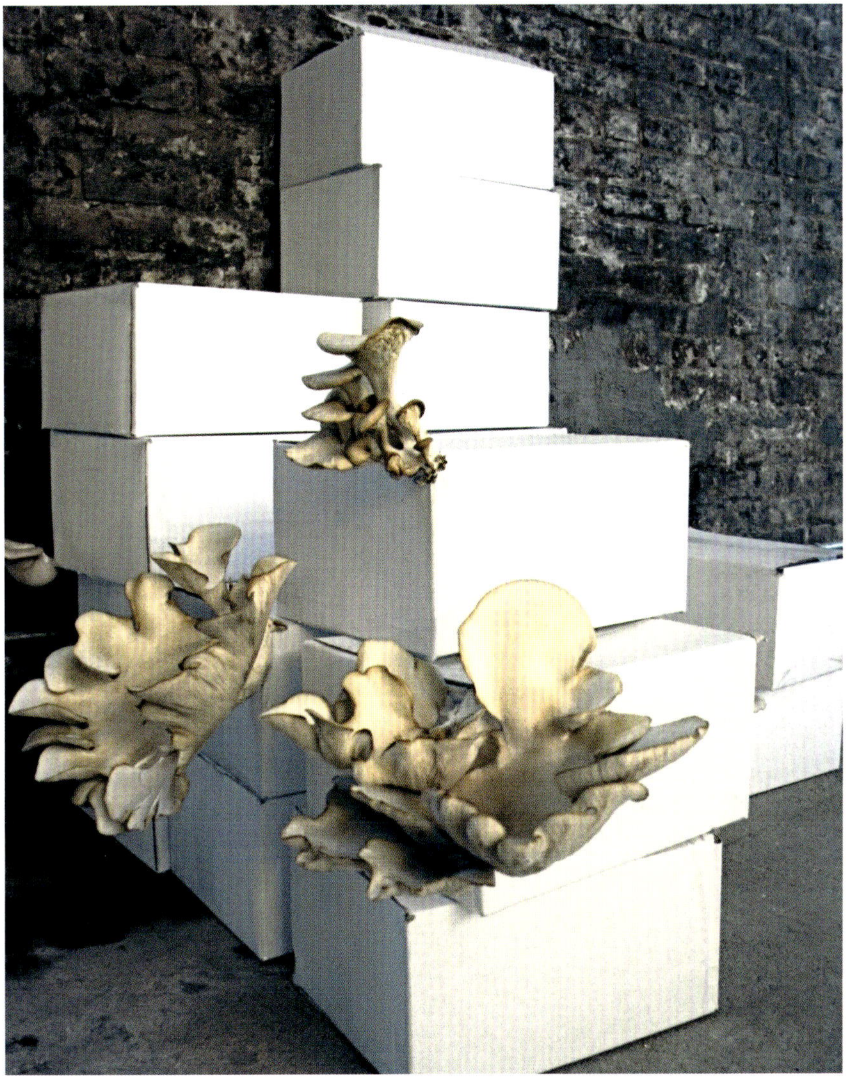

(this page and facing)
Win Me Free, 2005 (details)
Mixed media, live mushrooms, painted cardboard
Installation view, Silo, New York

IB Can you describe *Table of Contents*? Why did you make it?

PL **It came about while I was making my show, *Problem Picture Sources New Sculptures*. As I was doing all my making and painting and blurring, and touching the computer all the time and looking and reading, I was thinking, "What is holding all this stuff together? I'm trying to make a show here," and I realized it was the text. *It's like a text.* So I decided to make my own book. I made that book cover and photographed it. It's a model book. The thing on the cover of that book is from a painting I did of a lamassu [the Assyrian winged bull god with a human head] at the Met. That whole show has about five or six sculptures that I've been obsessed**

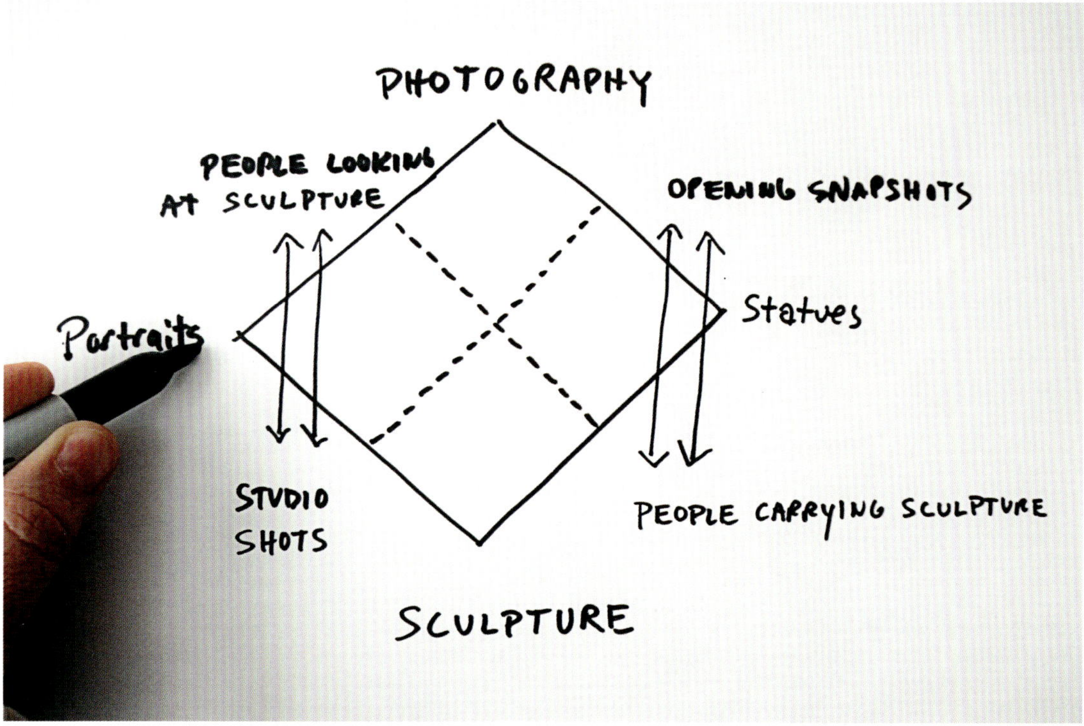

Klein Chart, 2010
Photograph

with for a really long time.

IB Historical sculptures.

PL **Yes, and classical sculptures.**

IB Brancusi, *The Lincoln Memorial*....

PL **Yes, also things in the Met from ancient Syria, Donatello, some Italian Renaissance stuff, Michelangelo's *Dying Slave*. There are a few that have been with me for a really long time. So I made the book.**

IB Made the book as an object.

PL **Right, because it's the text without a body. I decided to have a table of contents because it could give me a place to structure everything. So I created a structure, put in the thoughts of the show, and so the table of contents came about. It's really about how to look at the show, how to think about sculpture, and how to address some messy thinking.**

IB What is messy thinking?

PL **Thinking about the work and doing a show and articulating it means cramming a million ideas into one place at one time.**

IB Several artists I have talked to have an attraction to modes of not knowing, like a place that's messy or not fixed. I'm wondering if you are attracted to that, and what that offers.

PL **I think that whenever you sit and look back at the work you're no longer in the not-knowing. When I'm making the work, I think I let go of some ideas. But then you also have the history of your own work, which gets thrown in there, and the irrational part comes into it. I allow myself to work in the space of unknowing, or not knowing, for a while, but at some point there is a knowing.**

IB In *Owl* you reference an owl sculpture you saw at a church in France.

PL **Yes, it was on the edge of a medieval church in Dijon, France, and was carved by an anonymous sculptor. There are a lot of versions of the story, but during the years the church was built, there were problems with the working conditions—workers weren't being treated right and were getting hurt—and the owl became a token of good luck. It's a really beautiful thing because in the town everyone will walk by the owl and touch it. I really took pleasure in the whole procession of coming toward it, touching it, and walking away. It has to do with the idea we were talking about earlier about the optimistic future. I wanted to have that in my sculpture, so I made it.**

IB Your newest series combines armatures from Henry Moore and colors for Pontiac's 1974 GTO. They come from a specific time and a specific car.

PL **That is one way to frame it. 1974 was a big year: it was the last year the GTO was made and the year we pulled out of Vietnam.**

IB Is that how you got to that specific car and its colors?

PL **I became interested because I like to read about color, pigments, and paint applications. I'm in a color reading group that I would like to give a nod to. There was a point when there were a lot of reds because of car companies. I think it started in Europe with Fiat, and then it came over to America, and it has affected how people paint. They were the** people with the money to do the chemistry and get away from the pure pigments. The GTO was also a car that kicked up a lot of things with me, so it is probably some combination of fetishizing it, or at least conducting an irrational relationship with it, and claiming it. It's a classic American item. It's similar in some ways to claiming a Brancusi, but something I knew differently than from photographs. There is also the economy of the car industry in America which is a big foundation.

IB Why pair that with a Henry Moore unfinished work?

PL **An armature that I didn't even know was there.**

IB You're exposing the interior or the underneath.

PL **And suggesting the thing that's going to come.**

IB You mean the idea of an armature?

PL **It's all mysterious to me. It's like a subliminal foundation for something else. It doesn't look like its finished self. That interests me.**

IB That it doesn't look finished?

PL **Well just what an armature is. I'm not sure that Henry Moore armature was used for a finished piece. It might be an armature for a model. I don't always do a lot of research to find out what the thing is that I find. I find it and I consider it.**

IB What do you think about that notion of revealing your process? You've talked about your *Table of Contents* as a guide to your last show.

PL **I used it for many things: it's an art piece, it's a press release, it offers some thoughts, and it's an art statement.**

IB So viewers should have some information.

PL **There is always a text floating around somewhere—everything is so complicated.**

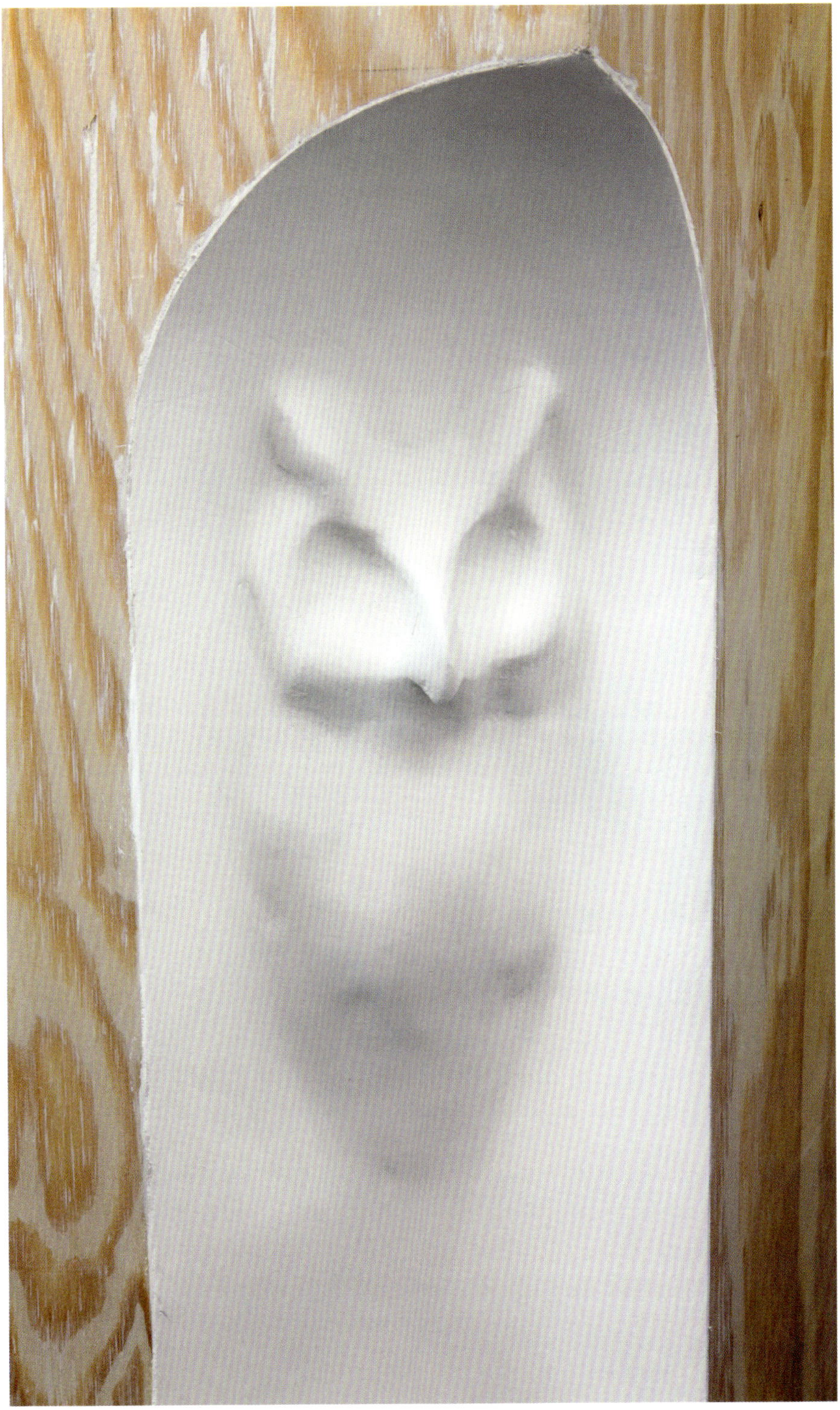

Installation view detail, *Owl*, Rachel Uffner Gallery, New York, 2009

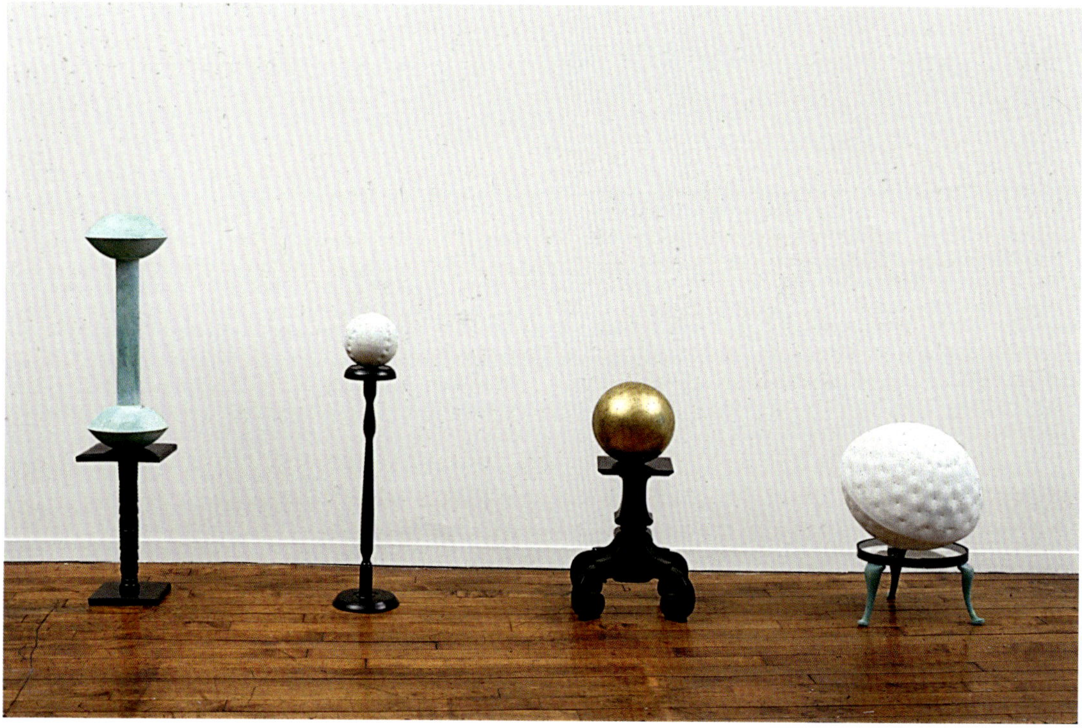

Sports, 1990
Patinated brass, metal, wood, color
Dimensions variable
Installation view, Art in General, New York

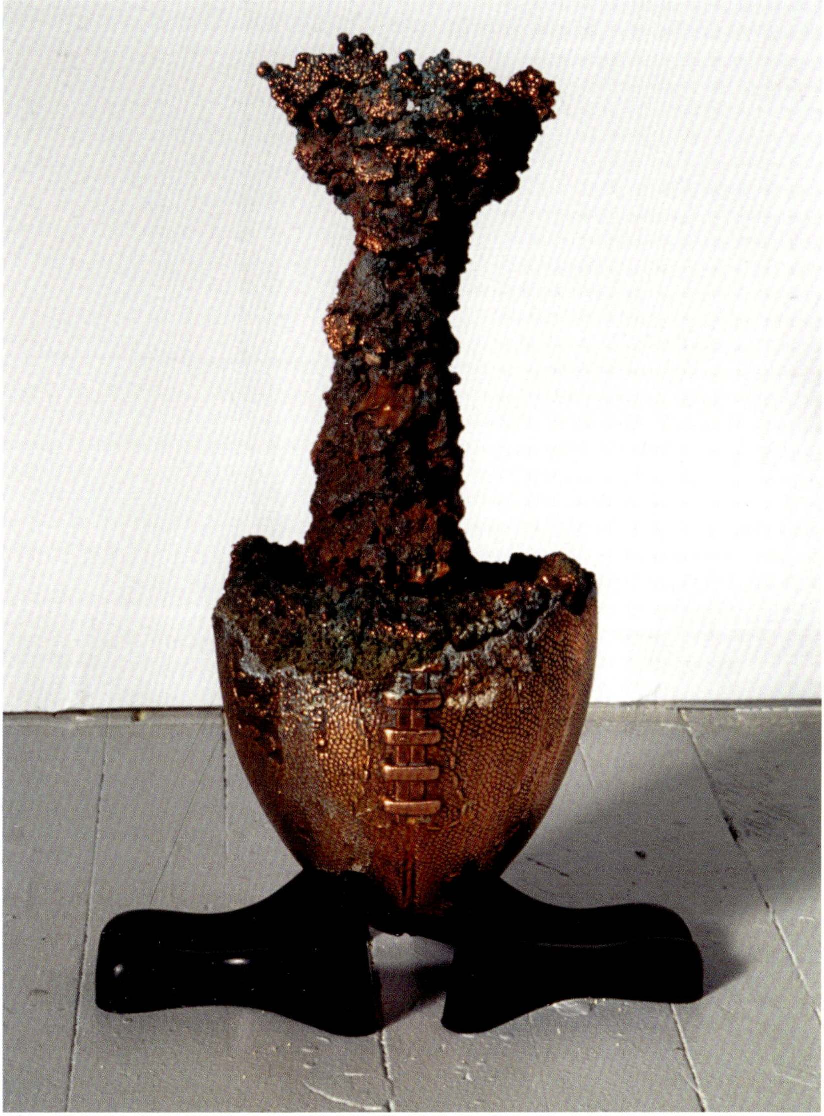

Deep Play, 1994–95
Electroformed nerf ball,
plastic
12 x 6 inches

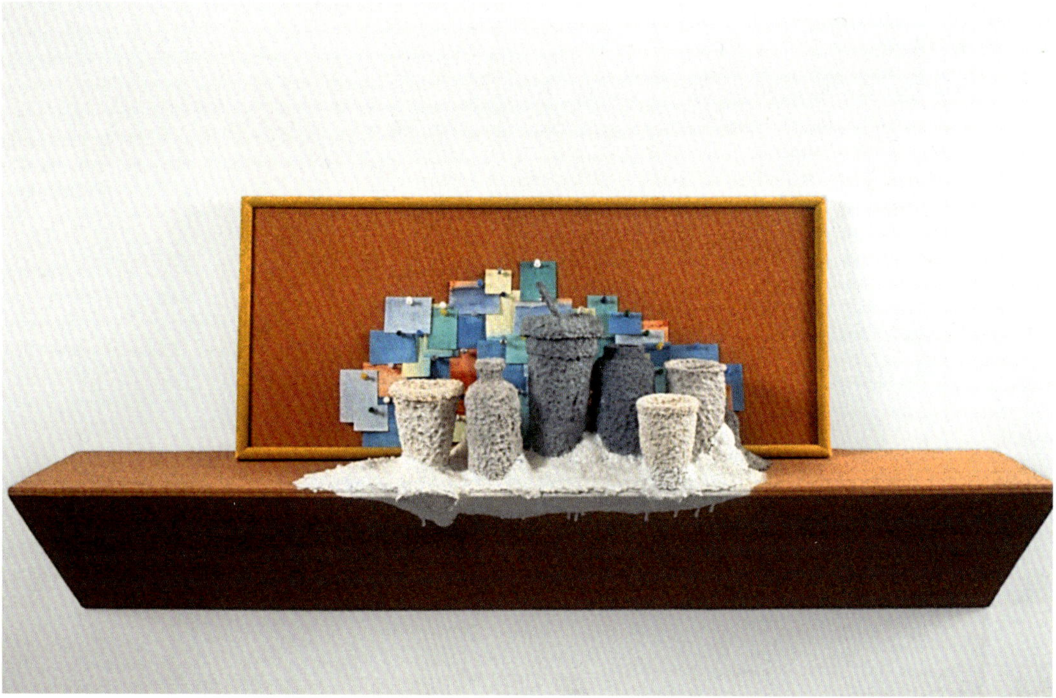

Sip Cup Memo Board, 1996
Foam, cork, wood, paper,
flocking
48 x 72 x 14 inches

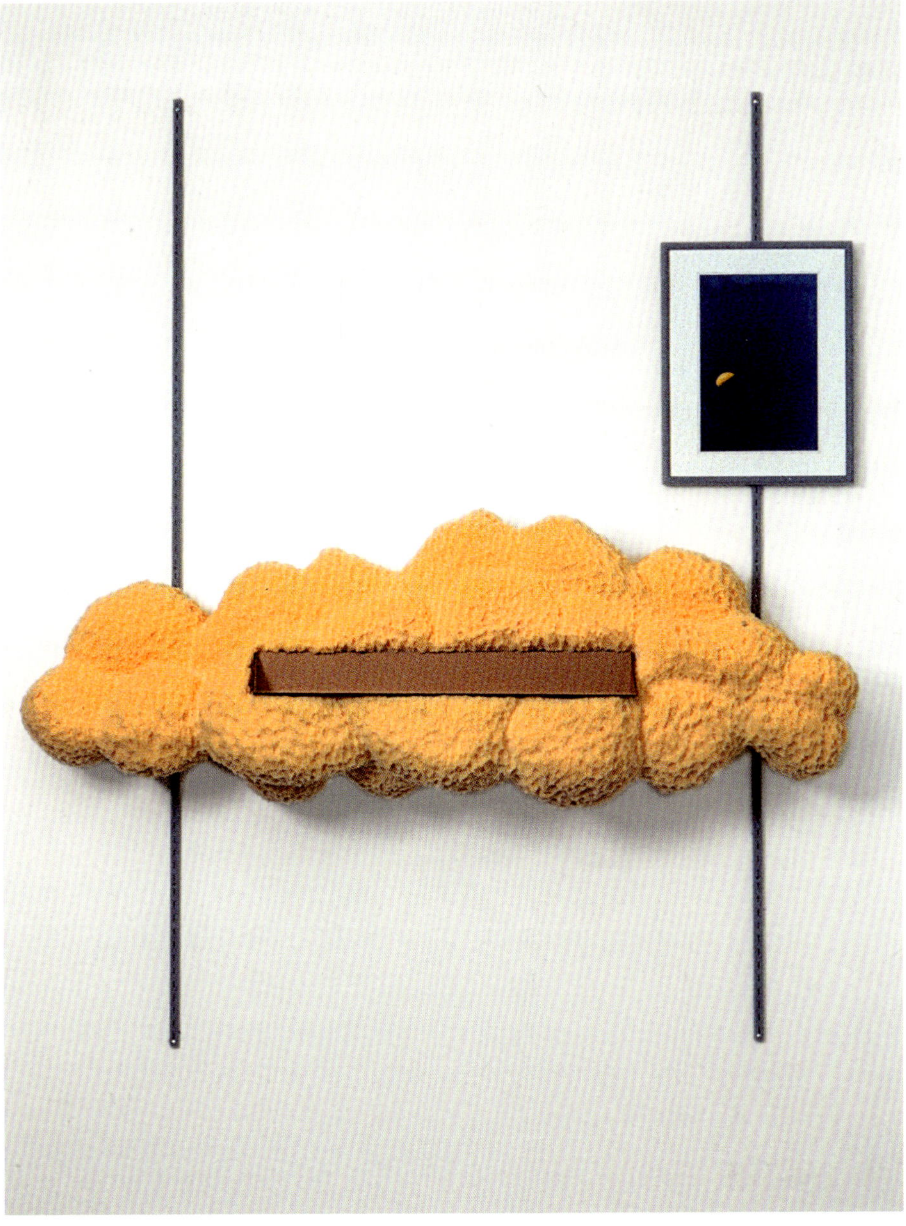

Cloud Shelf, 1996
Handpicked foam, cardboard, photograph, flocked hardware
60 x 60 x 20 inches

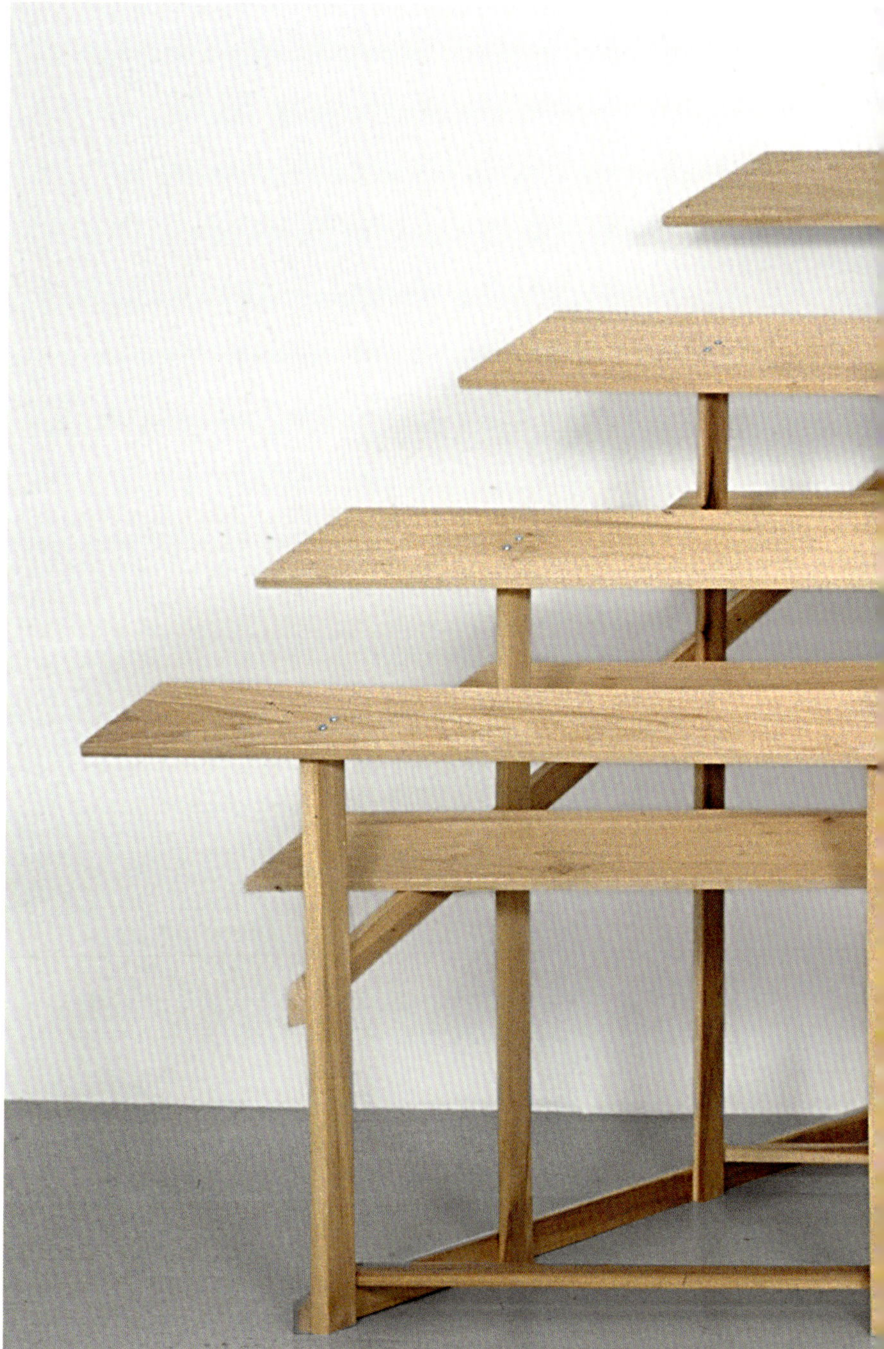

Bleachers (vent view), 2001
Poplar, hardware
62 x 104 x 50 inches

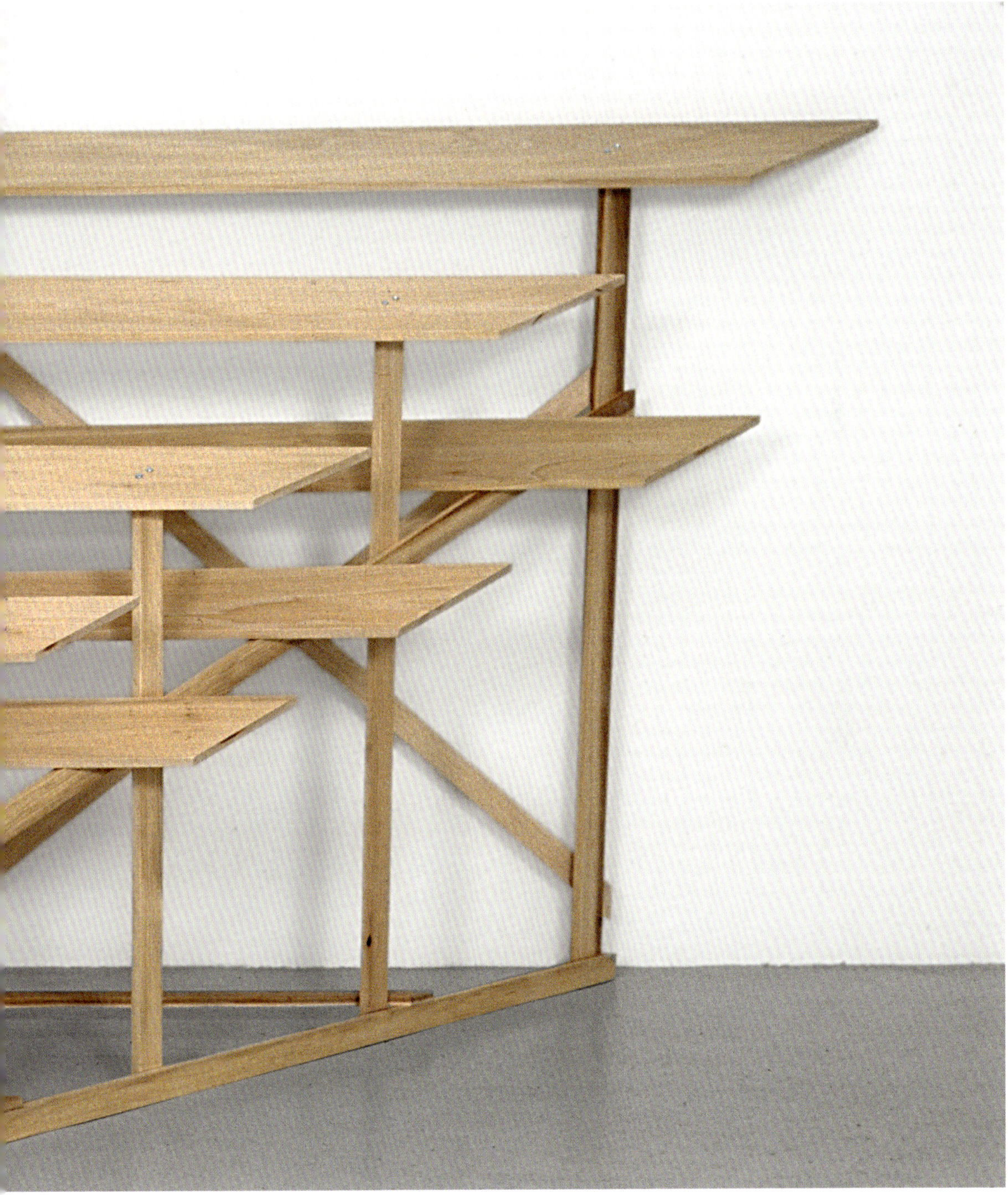

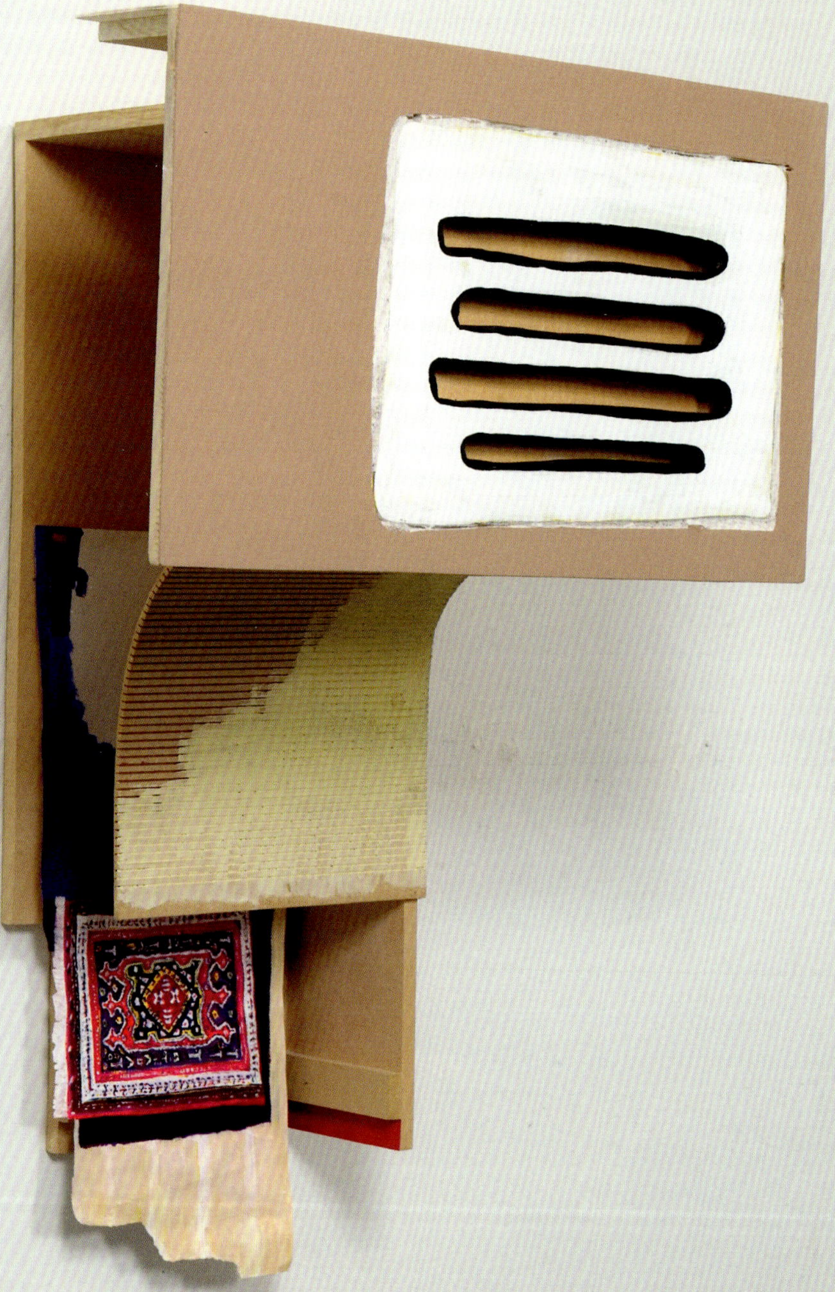

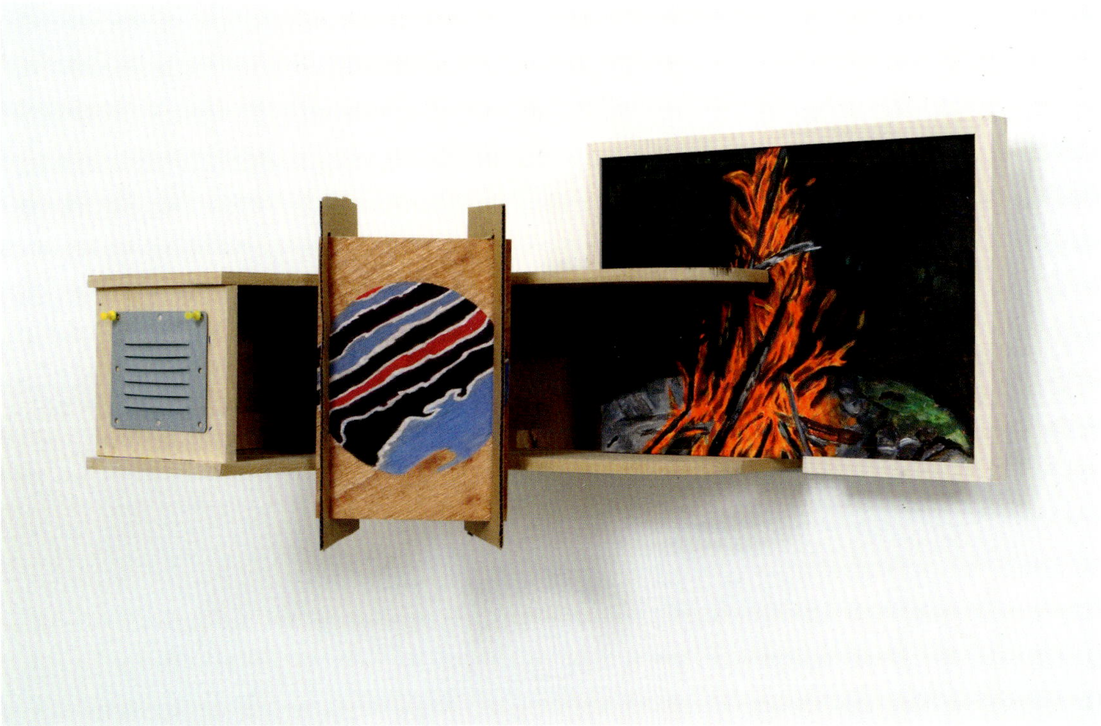

(facing)

Vented Rug, 2003
Wood, paint, acrylic on paper
36 x 28 x 24 inches

The Coast, 2003
Wood, cardboard, paint, acrylic
on canvas, metal vent
29 x 19 x 12 inches

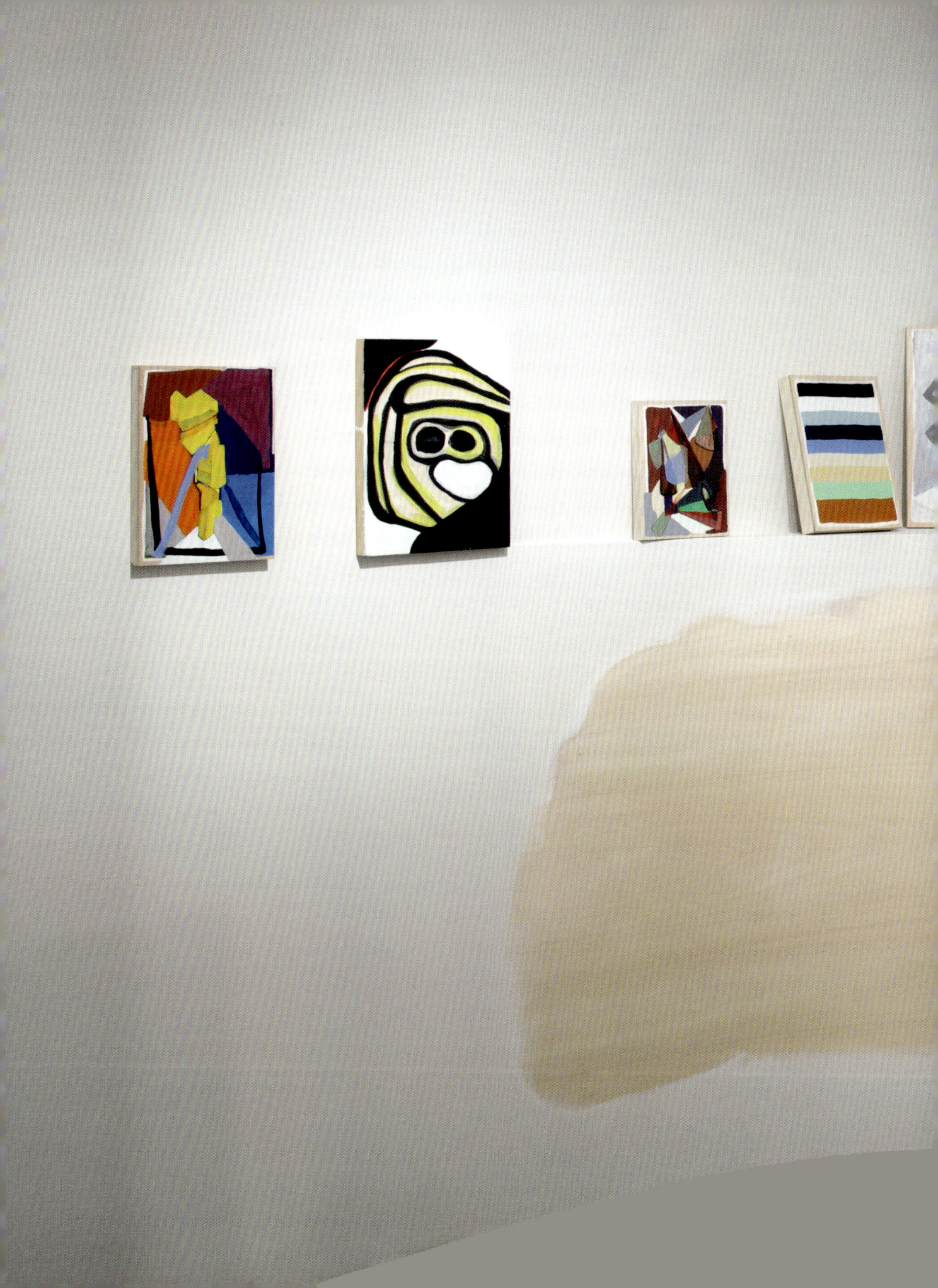

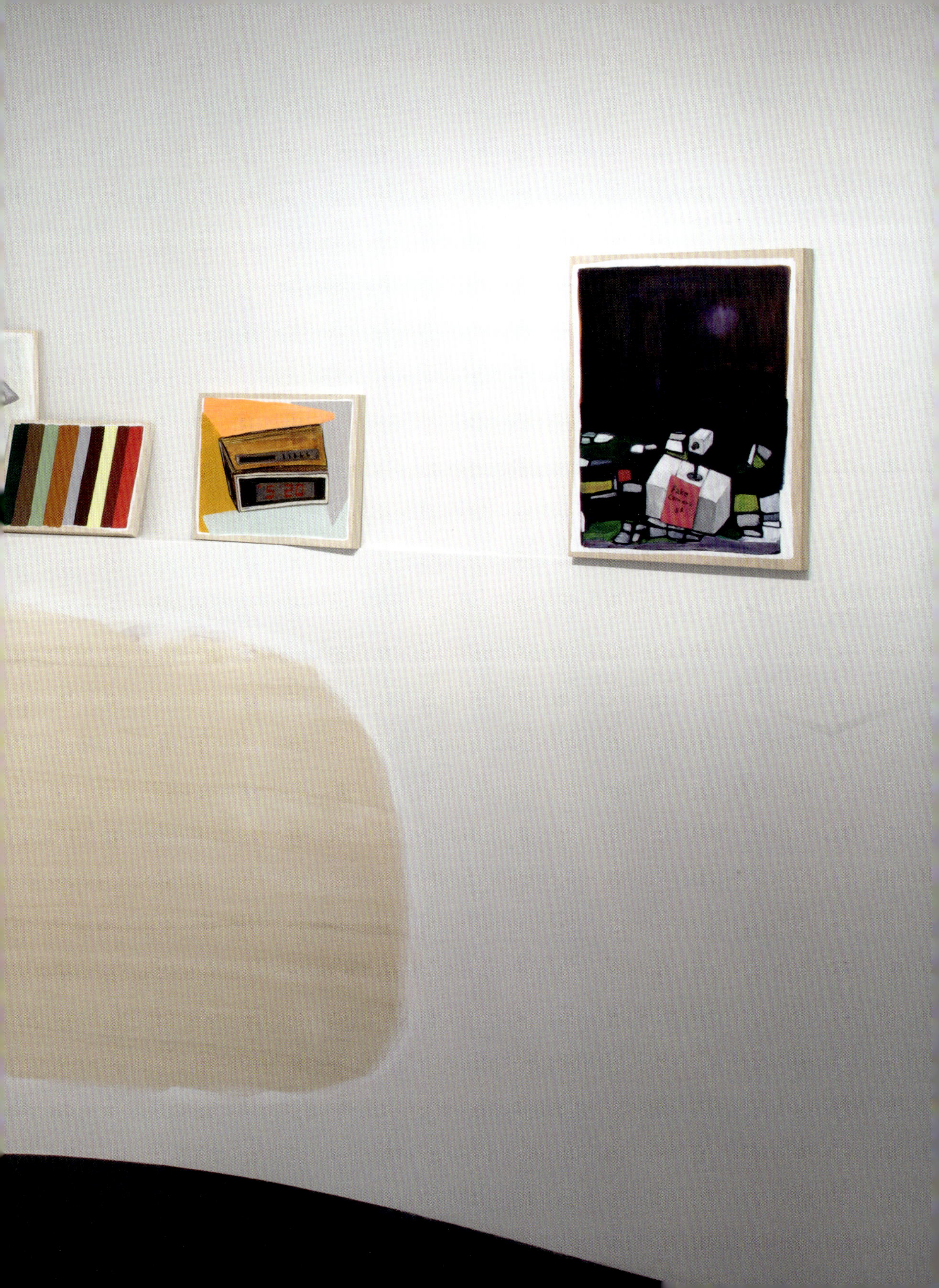
Fake

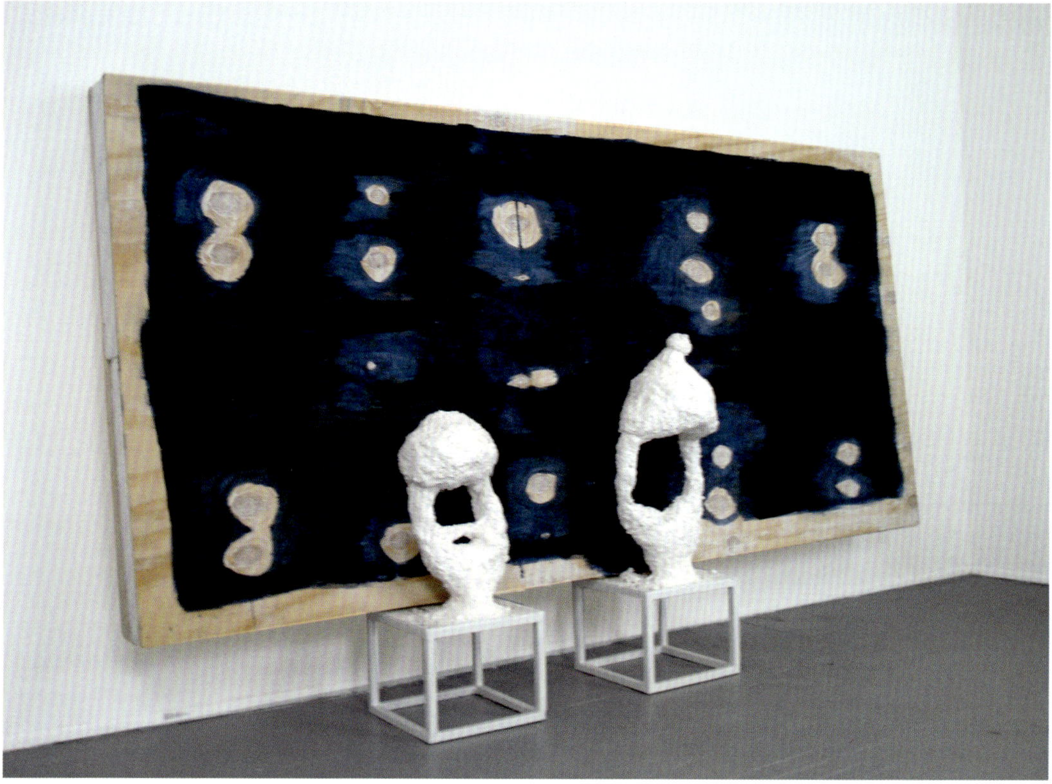

(previous spread)

Corner piece reconstructed painting background, 2003/2012
Plywood, paint, paintings
Dimensions variable, approx. 4 x 8 feet
Installation view, Tang Museum, Saratoga Springs, New York, 2012

(above)

Please Bear With Us, 2007
Wood, plaster, acrylic
Dimensions approx. 36 x 36 x 36 inches

(facing)

Please Bear With Us, 2007
Wood, plaster, acrylic
Dimensions variable

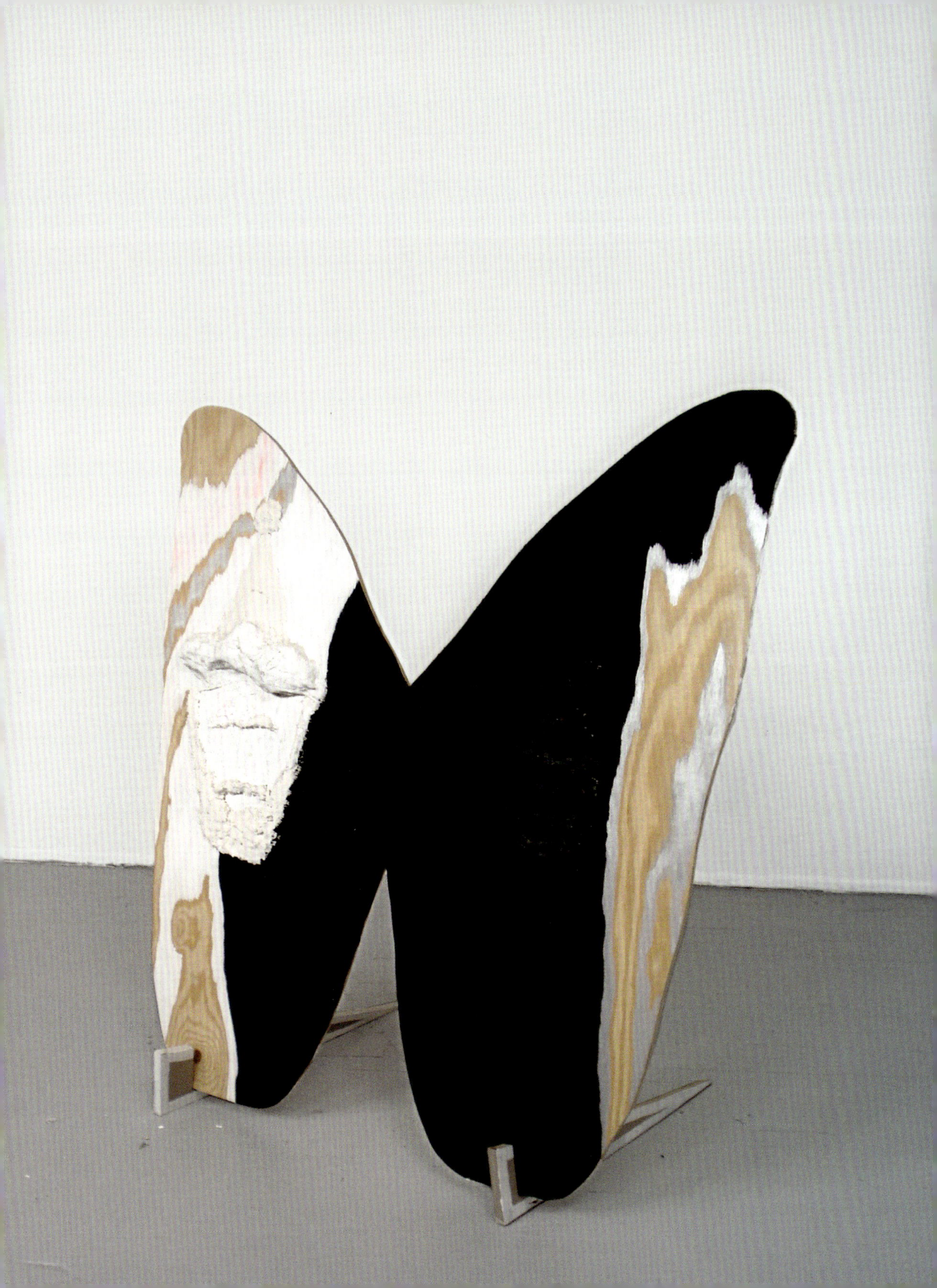

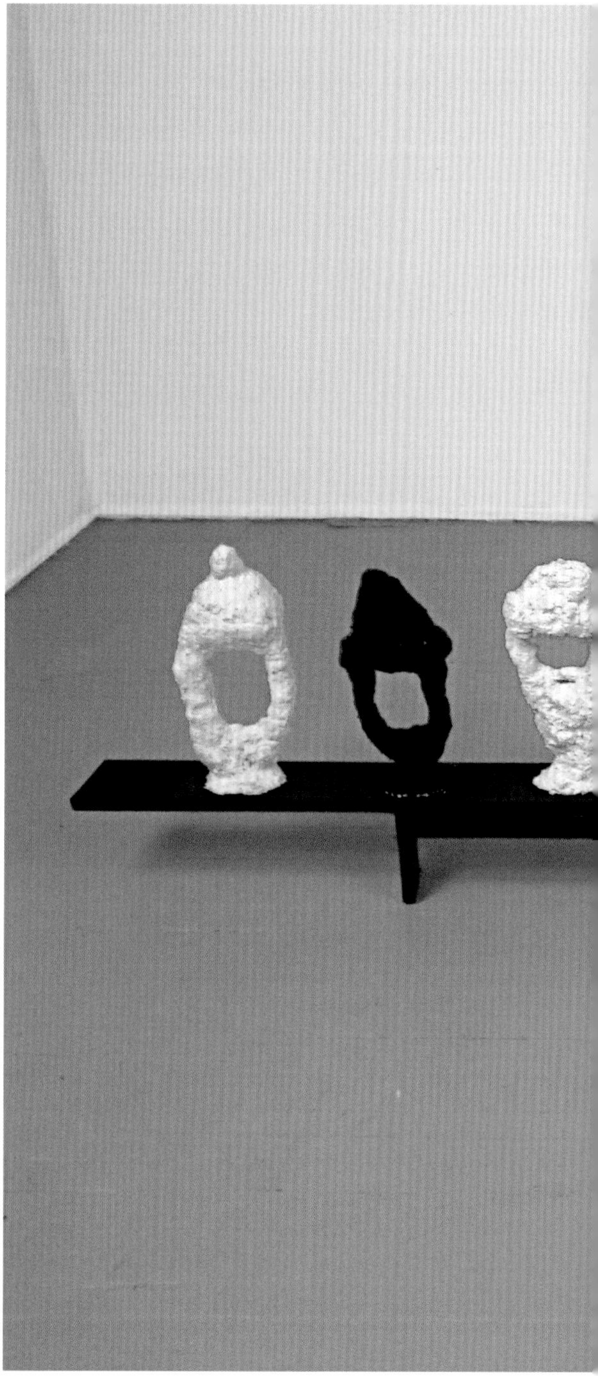

Please Bear With Us, 2007
Wood, plaster, acrylic
Dimensions variable

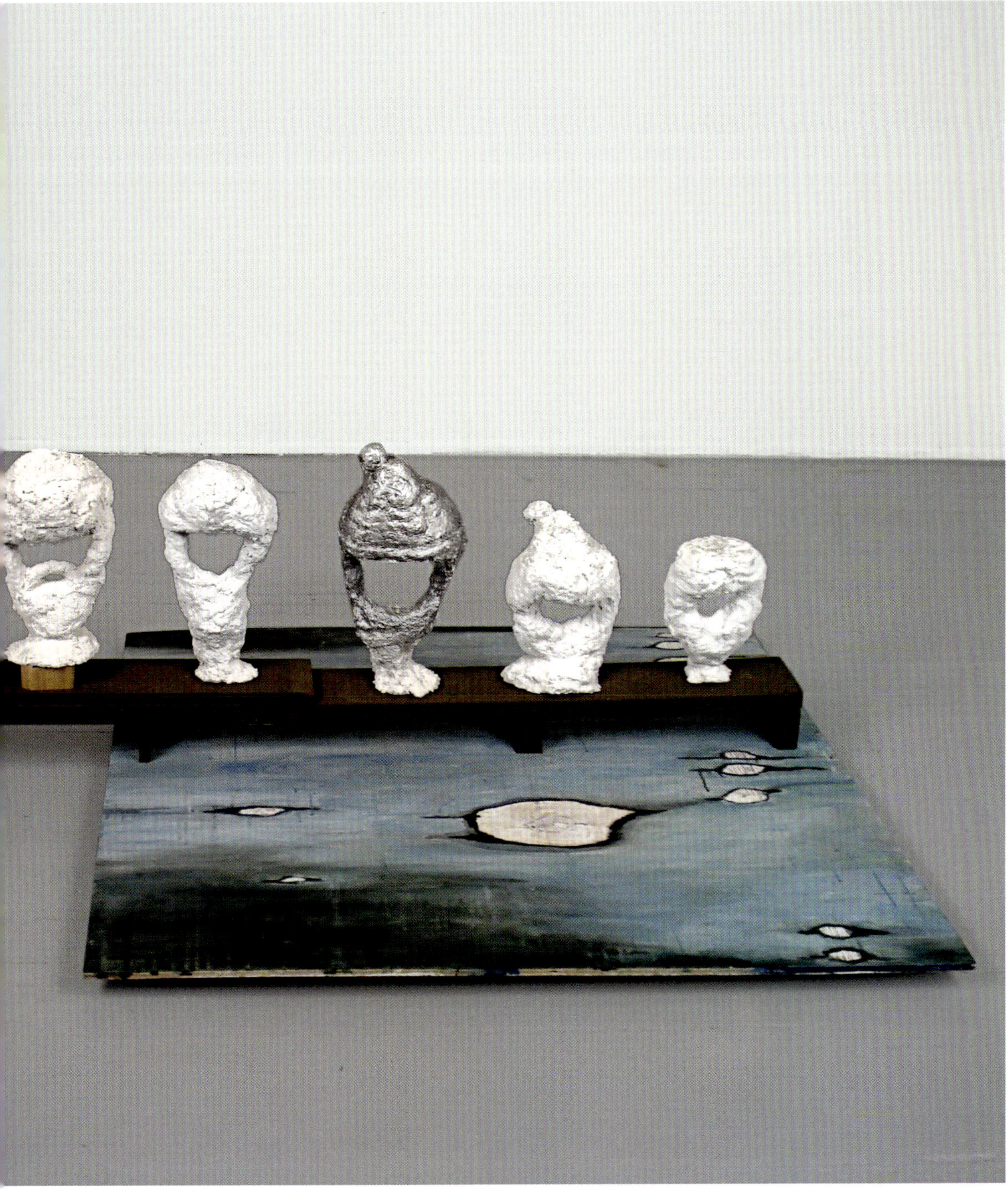

Installation view, *Owl*,
Rachel Uffner Gallery,
New York, 2009

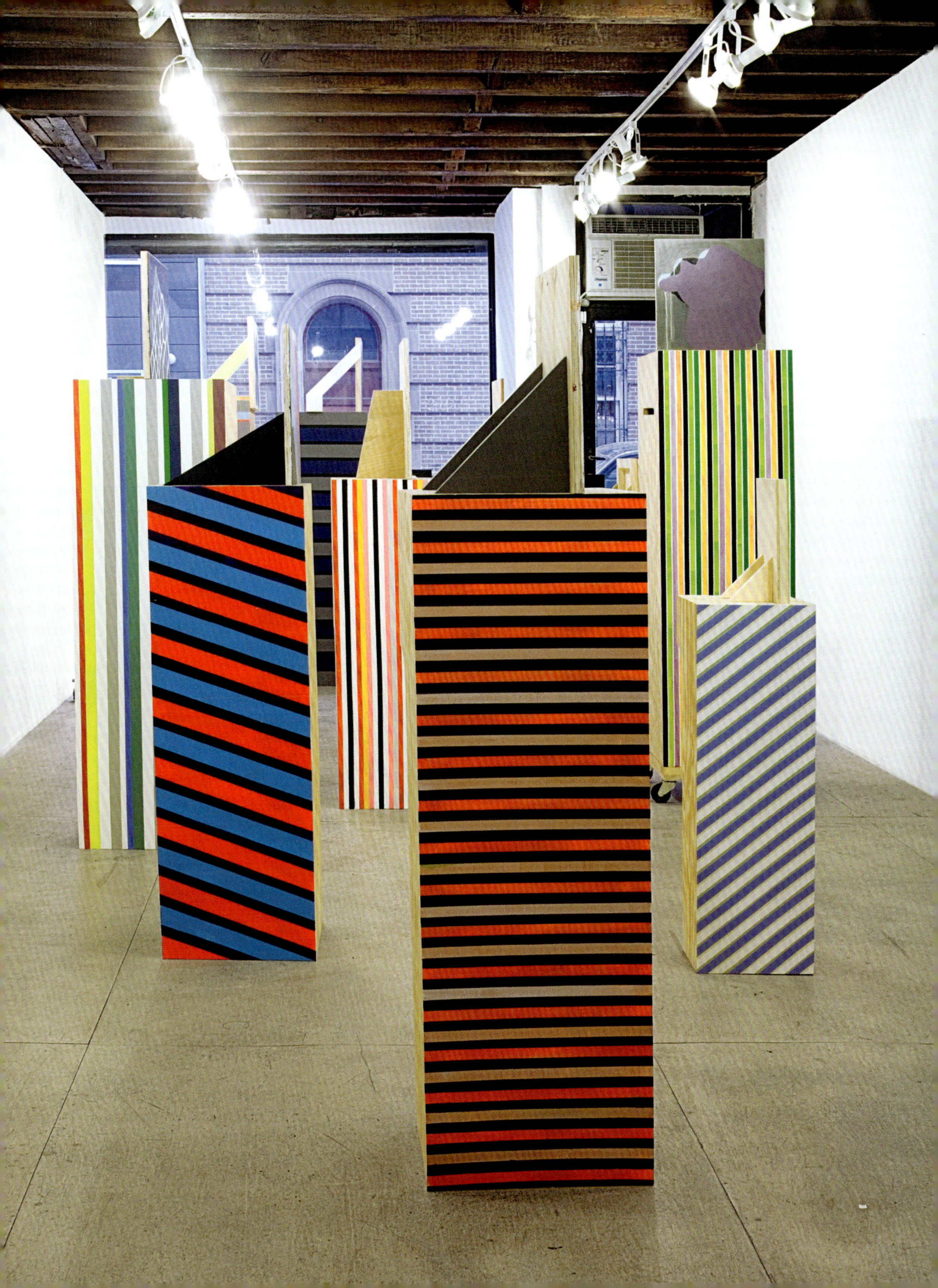

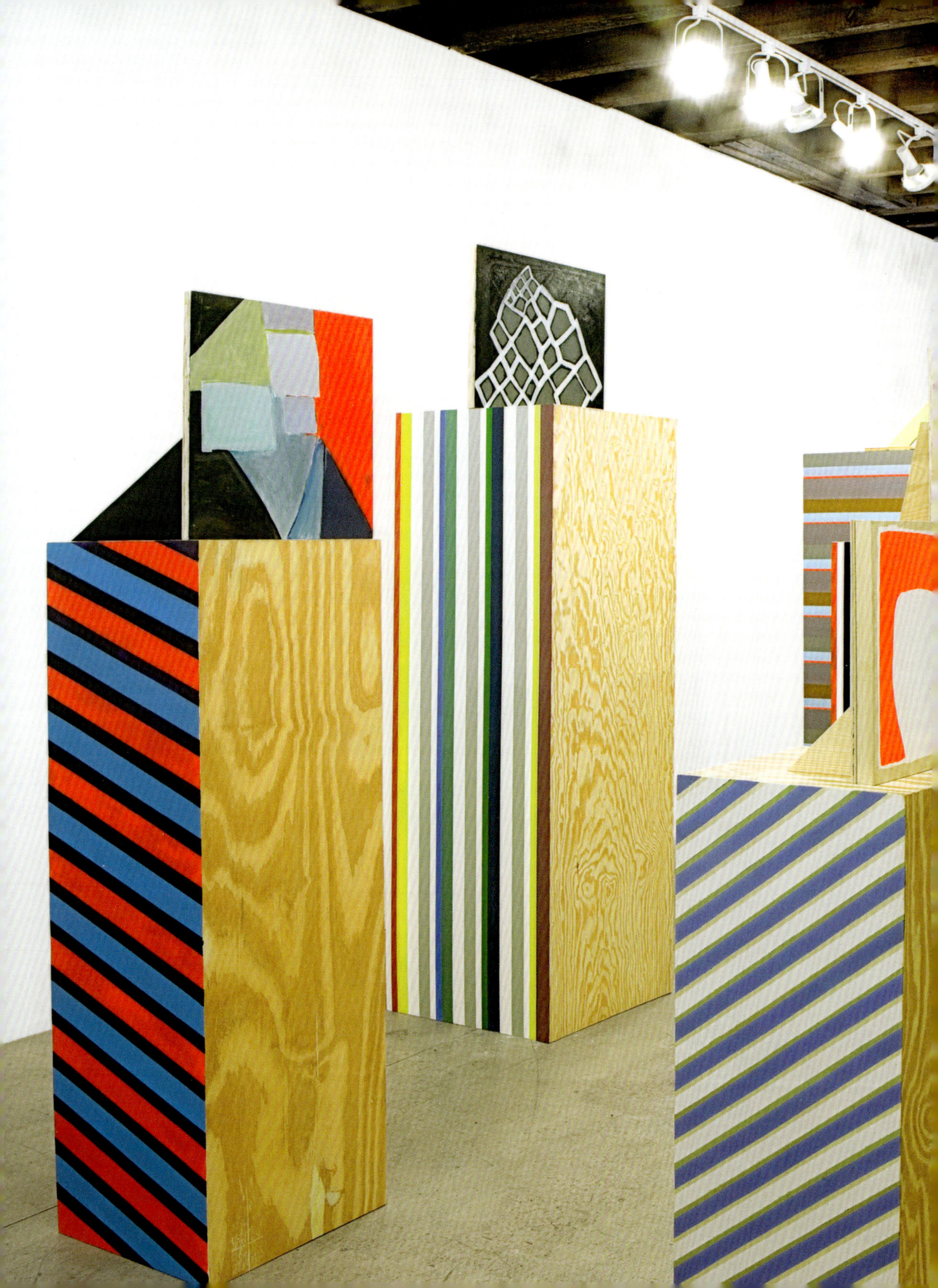

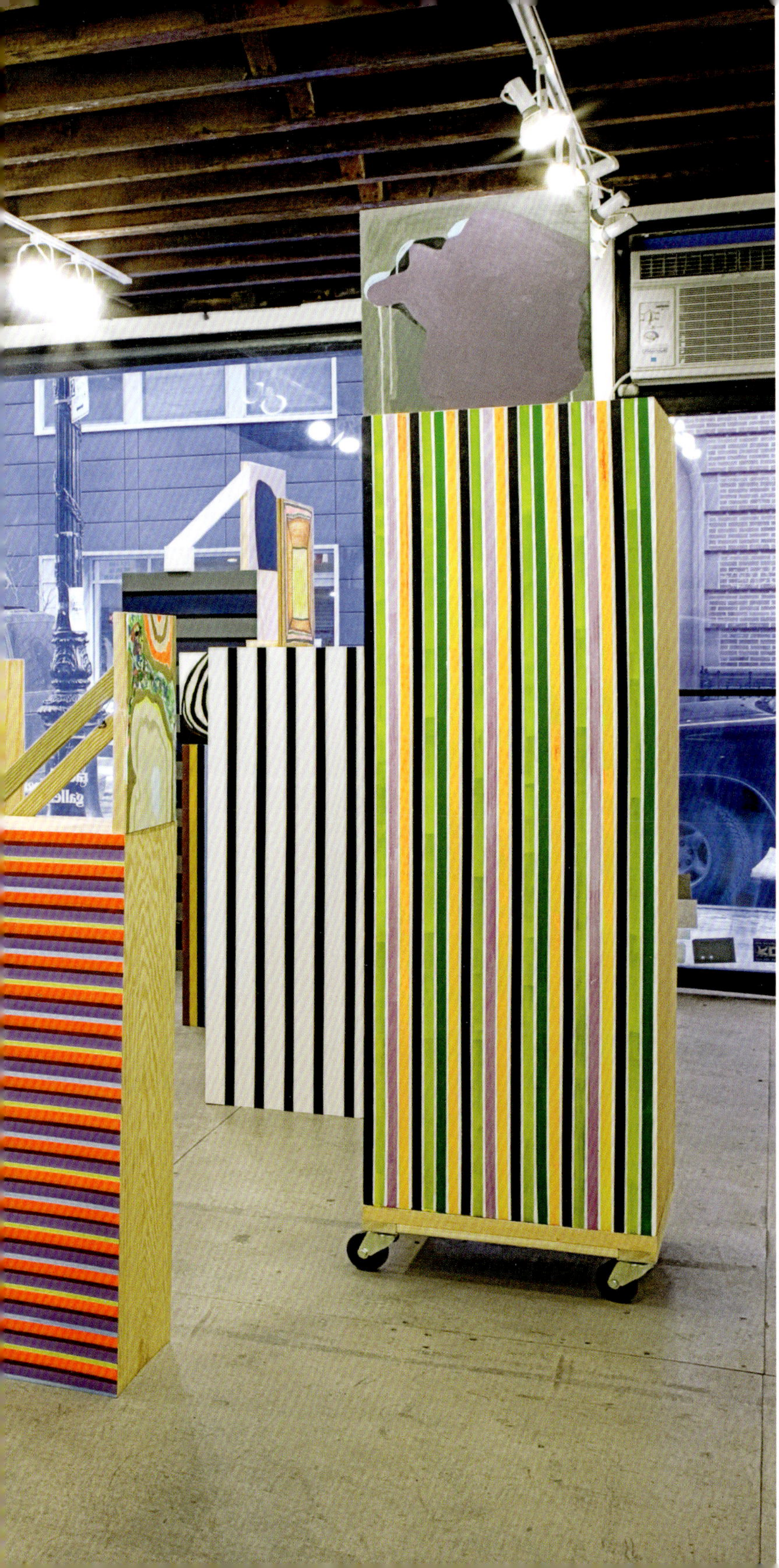

Installation view, *Owl*,
Rachel Uffner Gallery,
New York, 2009

Installation view, *Owl*,
Rachel Uffner Gallery,
New York, 2009

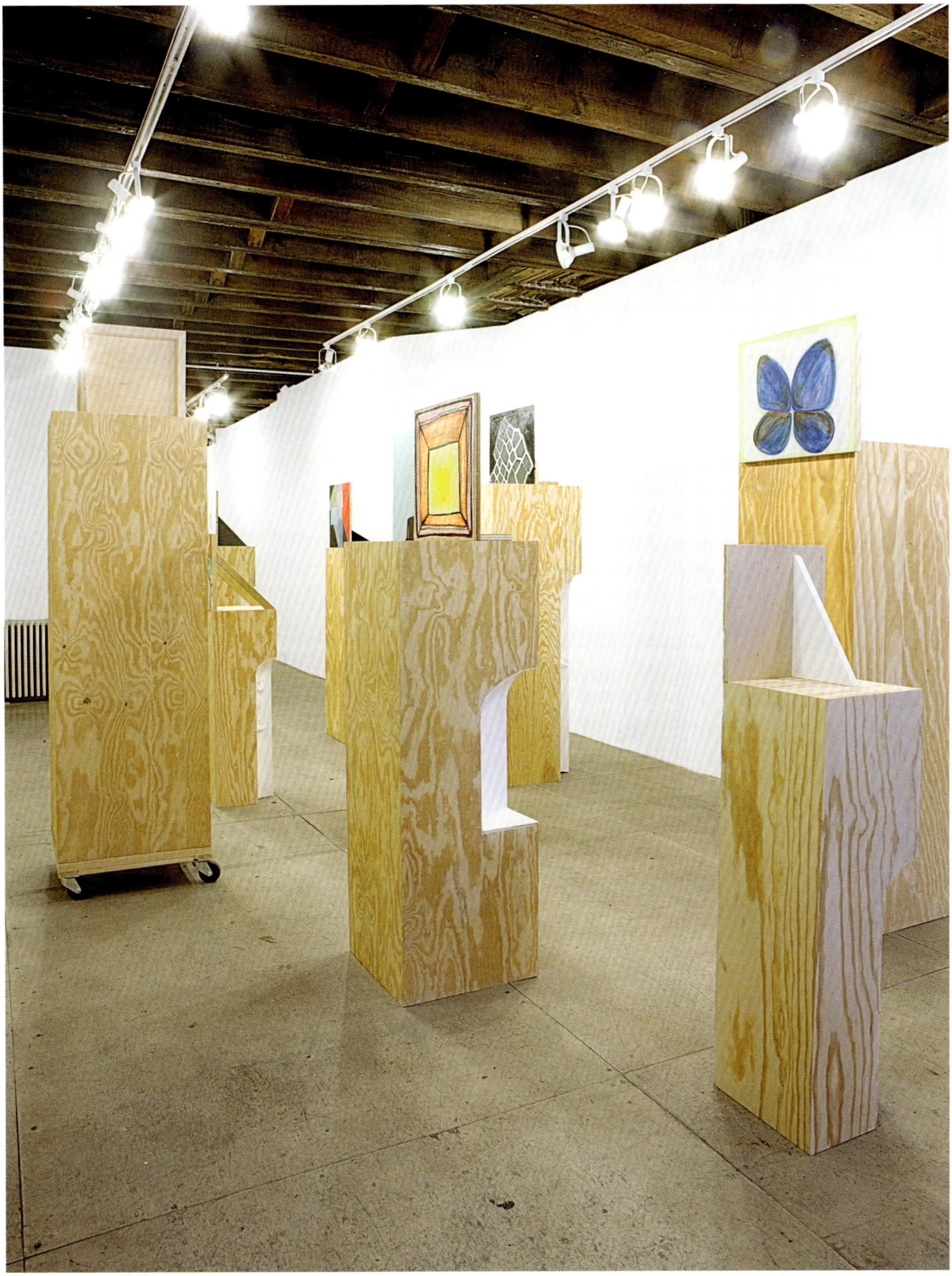

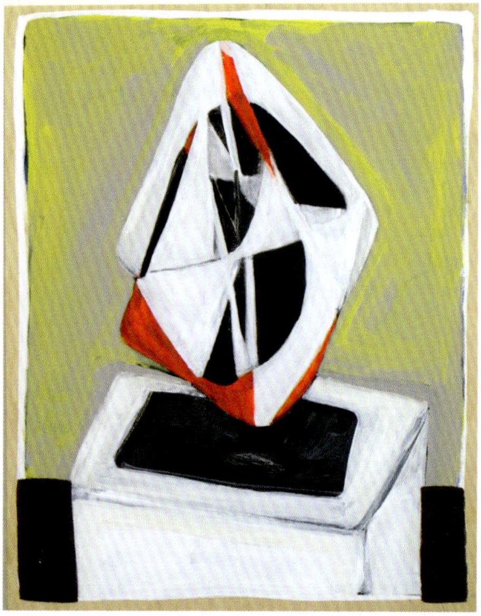

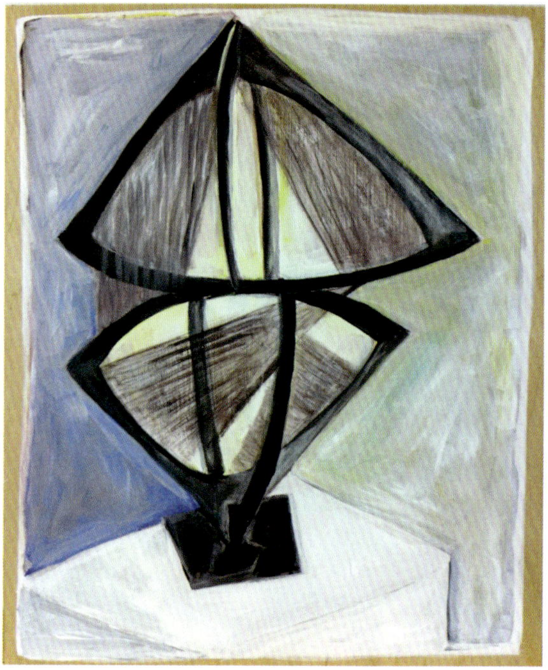

(left)

Gabo Delay #1, 2011
Acrylic on wood panel
18 x 14 inches

(right)

Gabo Delay #5, 2011
Acrylic on panel
20 x 16 inches

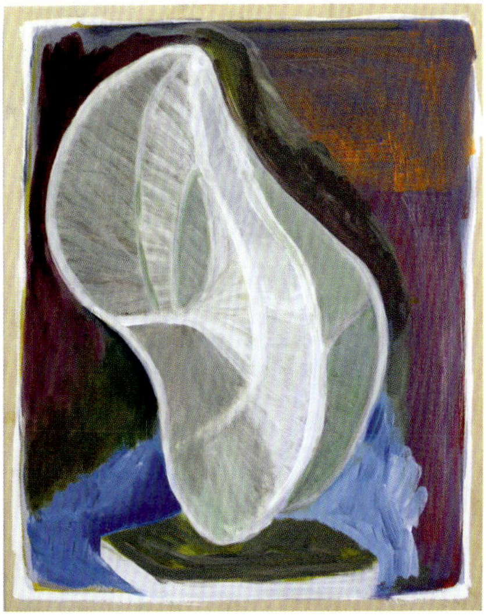

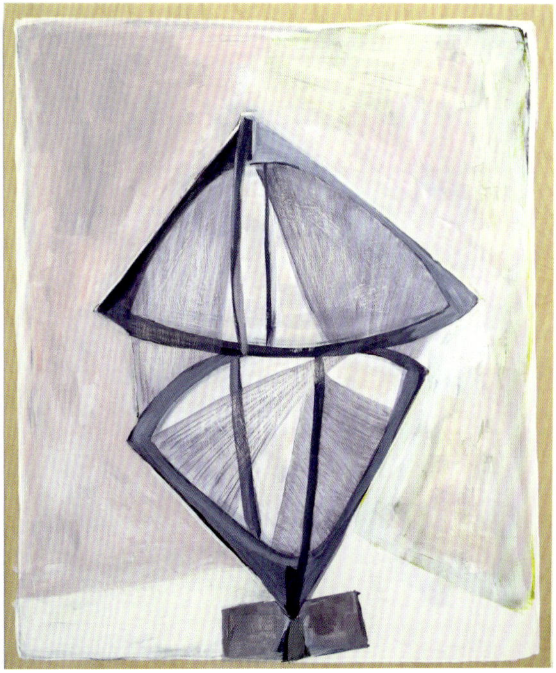

(left)

Gabo Delay #3, 2011
Oil on panel
14 x 11 inches

(right)

Gabo Delay #2, 2011
Oil on panel
20 x 16 inches

CONTENTS

PREFACE

1997—Brancusi's studio was taken apart and reproduced by Renzo Piano at the Pompidou Center in Paris. The studio is positioned behind unglazed windows. There is no physical access to the space, giving the visitor an analytical representation, but no consideration of distance. **QUOTING** The reinstallment of Brancusi's studio or/ Looking at his studio from the viewing window as a way to think about sculpture and photography. Moving the studio to the gallery. Reading the fireplace in Brancusi's studio as the threshold of sculpture. The fireplace as an empty niche, Brancusi's *fond de toiles* as a way to look at sculpture. **THE STUDIO—THE RESEARCH—THE WORK** The psychological spaces of practice. Viewing and making, the subject/object locale. The role of the text as keeping it all together. The text without a body. Visual representation is made up of legible signs. Representation that loses its' clarity and defies rational understanding may support cognition and liberation. Everything lost or disappearing needs a monument. Using materials as they are vanishing. Driving in cars enables the near past to disappear at the same time it is reflected on it's body. The operative themes of the work move on and are replaced by other formations. The sculptures as absorbers of the passer-by. Emptiness in the act of depiction. Considering the dialectical relationship of photography and sculpture; double blindness. The classic forms of sculpture and the value of the newly rejected. Obscuring a sculpture with a painting. Never leaving, giving access to the archive.

INTRODUCTION

EMPTY NICHES / GHOSTS MIRRORS / GTOS / DECOYS / DELAYS / OBSTRUCTIONS

CHAPTERS

REPRESENTING THE BODY Carving the likeness of others, to sculpt or depict. The slippage of representation. Resemblance of life and how a statue teaches us about loss. Looking at the eyes, sculpture looks back. The gaps between carved hair, missing teeth, and extra legs. Veins, moles, wrinkles, and bumps: reading the neurotic details. The sunken eye vs a protruding eye.

REPRESENTING THE FABRIC Ruined fabric and veils. The wizard's robe; carved fabric, fabric fabric, tears, creases, folds, and pleats. The wet t-shirt and mimetic cloth.

REPRESENTING THE PROCESS Carving moments into spaces. Aggressive mistakes. Notice to the unbuilt. Polychromed sculpture missing the original paint. Paint fading, disintegrating, and flaking, and the reconstruction of the whole. Virtual environmentalism and the imagination. On accidents, the unfinished and the undersides. Flattening and shattering.

REPRESENTING THE MATERIAL Reading the gestalt of the material. The natural and how we see entire objects. Fake clocks, slow clocks, and faults: how to read time and the beauty of imperfections. The difference between cracks, chips, and dents. Hacking away at mountains.

REPRESENTING THE STANCE Sticking out and sinking in; trees, logs, appendages, and how they act as conduits within the sculpture. Decapitated heads on bases. Liberation and weight. Embarrassed pedestals. Sampling body language.

THE SCULPTURES

THE LAMASSU CARVING THE 5TH LEG AND THE POV Creating a psychological and unnamed space. Composite figures for composite spaces. Funny things about opposing perspectives. Rushing towards a moment, letting the statue stand free. Getting down from the wall.

THE DYING SLAVE UNFINISHED AND MISPLACED Everything that presents itself is forgotten. Substitutes and contradictions. The creamy visage of resignation. Nothing is more visible than what is unbuilt. Flesh and sympathy and crisis.

MARY MAGDELENE THE CHANGING FACE OF THE FEMALE FIGURE Resemblance as drama. Crying styles. A figured figure, and a figuring figure. There are no symptoms without the act of forgetting. From starved aesthete to desire: power / starvation / depiction / depth. Hysteria in sculpture. Dark circles, anorexia, and the flesh of the future. Where figuring is equal to disfiguring. Donatello's *Mary Magdalen* and the women in Antonioni's films.

MEDUSA SHOWING FEAR OF THE SELF Sculpture regarding us and itself. Following the pathic dimension of the subject. Seeing before knowing. The viewer as reflecting pool. The carving of a symptomatic moment. The ghost mirror. The uncarved eye. Staring into a space with a void. Removing the object of focus.

THE LINCOLN MEMORIAL THE BOOKEND Robert E. Lee and the back of the head: carrying his nemesis within his representation. The past screens itself. How to use space to reinterpret beliefs. The invisibility of monuments. The buttons on the sleeve and the psychology of scale. The death drive and work of the memorial.

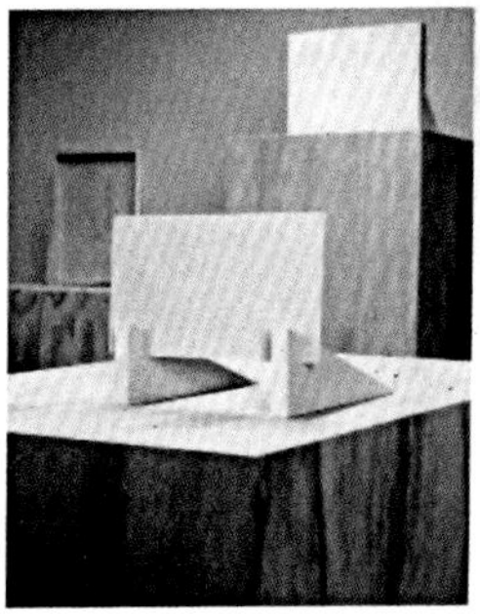

AFTERWORD: SIX QUESTIONS

How to see the sculpture from the place it never sees you. Shiny surfaces and the Midwest.
How sculpture tells a joke. What we learned from Buster Keaton and Tati.
How to not touch the touchable. Viewing and making and how to resist the haptic.
How to stand next to a sculpture. Being alone or being together.
How to live with a sculpture. Reading the architectural spaces around a sculpture. (the free zones)
How to pretend to be a sculpture. The uncomfortable openness of suspension.

NOTES

ESTABLISHING A RELATIONSHIP TO PHOTOGRAPHS AND SCULPTURE **1.** The opening and closing of space: photos of sculptures in a book. **2.** Photography as the primary medium that figurative sculpture has fought against. **3.** The sculpture photo as end or opening. **4.** Using a photo to capture a look, but not to historicize it. **5.** Repetition and the case for Flickr. **6.** Problem picture sources, using the photo to minimize the effect of scale. **7.** Photos of sculptures and sculptures are not contradictory forms. When a photo draws on a system of perspective in order to produce 3D space, it opens directly into the world. **8.** The power of the negative space: looking into a photo and beyond. **9.** The shared crisis areas of photography and sculpture: the narrative and the fully static. **10.** Walking around a sculpture and running into a photo. **11.** Painting as an assist to the relationship to photography and sculpture. **12.** Tool trails, author residues, and flash marks, quoting quoting (reconsidering mark-making).

LISTS OF PLATES

Statue of a Kouros [fake?] at the Getty Museum/Flickr. *The Lincoln Memorial* being built/Records of the Office of Public Buildings and Public Parks of the National Capital. Stalin statue falling, Hungary, 1956/anonymous. Two British soldiers carrying a statue from Tut's Tomb excavation/Harry Burton archive at the British Museum. *Life Magazine* cover of *Venus de Milo*, Gjon Mill/Getty Images. Rapheal Monti, *Veiled Vestal Virgin*/still from the movie *Pride and Prejudice*. Robert Mapplethorpe, *Slave*, rephotographed with the artist's hand, gift to the Guggenheim Museum from The Mapplethorpe Foundation c/o the artist. Michealangelo, *The Dying Slave*, Clarence Kennedy archive/Harvard University. Donatello, *Mary Magdalen*/Flickr. *Portrait of Aymard-Jean de Nicolay*/Kimball Art Museum archive. Bernini, *Medusa*/The Capitoline Museum archive. *Bust of Lauren*, rephotographed, from *History of Sculpture—From Antiquity to the Present Day*, c/o of the artist. Photo of Joseph Stalin wearing a huge concrete coat, Margrate Bourke-White, Time and Life/Getty Images. *Grief*, by August Saint-Gaudens, Thomas Macavoy/Getty Images. Brancusi's rebuilt studio in Paris, commission by Rem Koolhaas/Flickr. B&w photo of Brancusi's original studio, by Brancusi/ Pompidou archive.

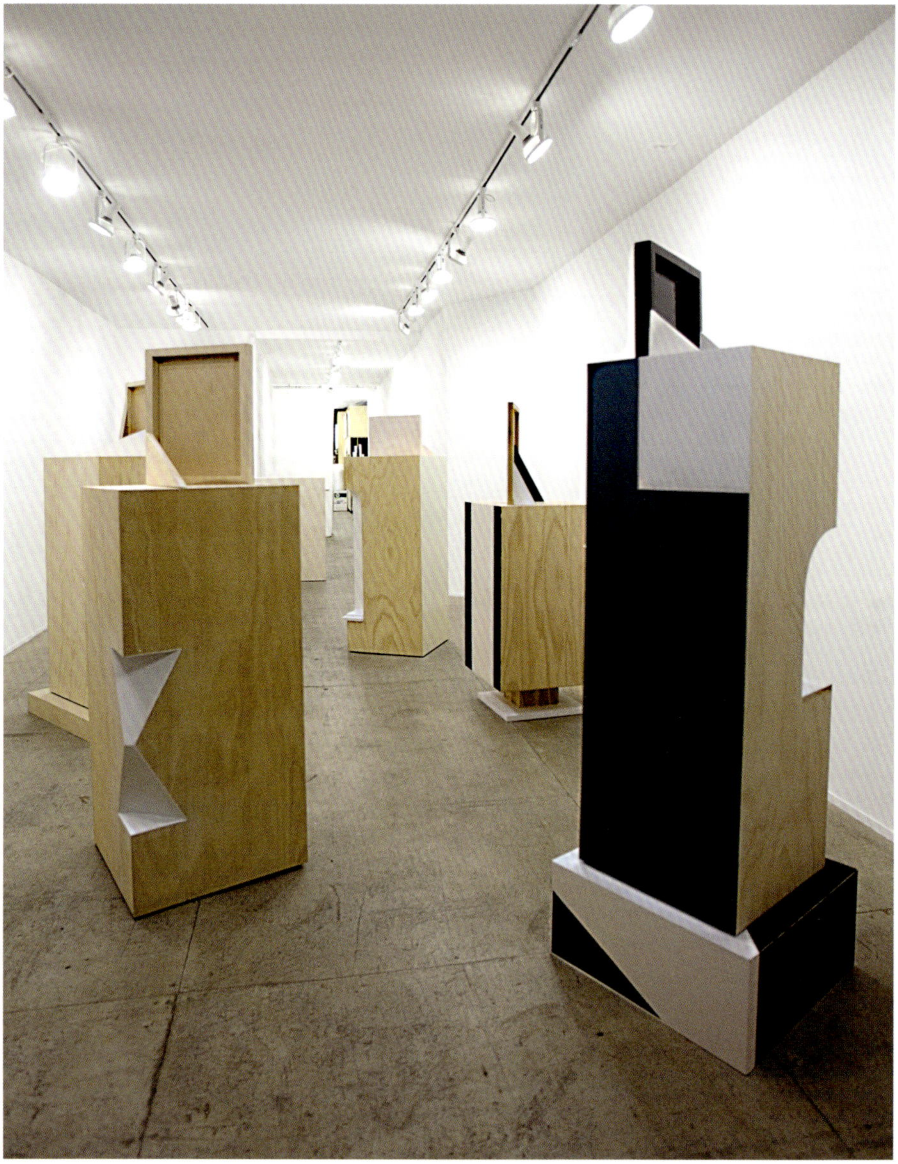

(previous spread)

Table of Contents, 2010
Archival inkjet print
12 x 18 inches

(above)

Installation view, *Problem Picture Sources New Sculptures*, Rachel Uffner Gallery, New York, 2010

(facing)

The Fifth Leg, 2010
Archival inkjet print
16 x 12 inches

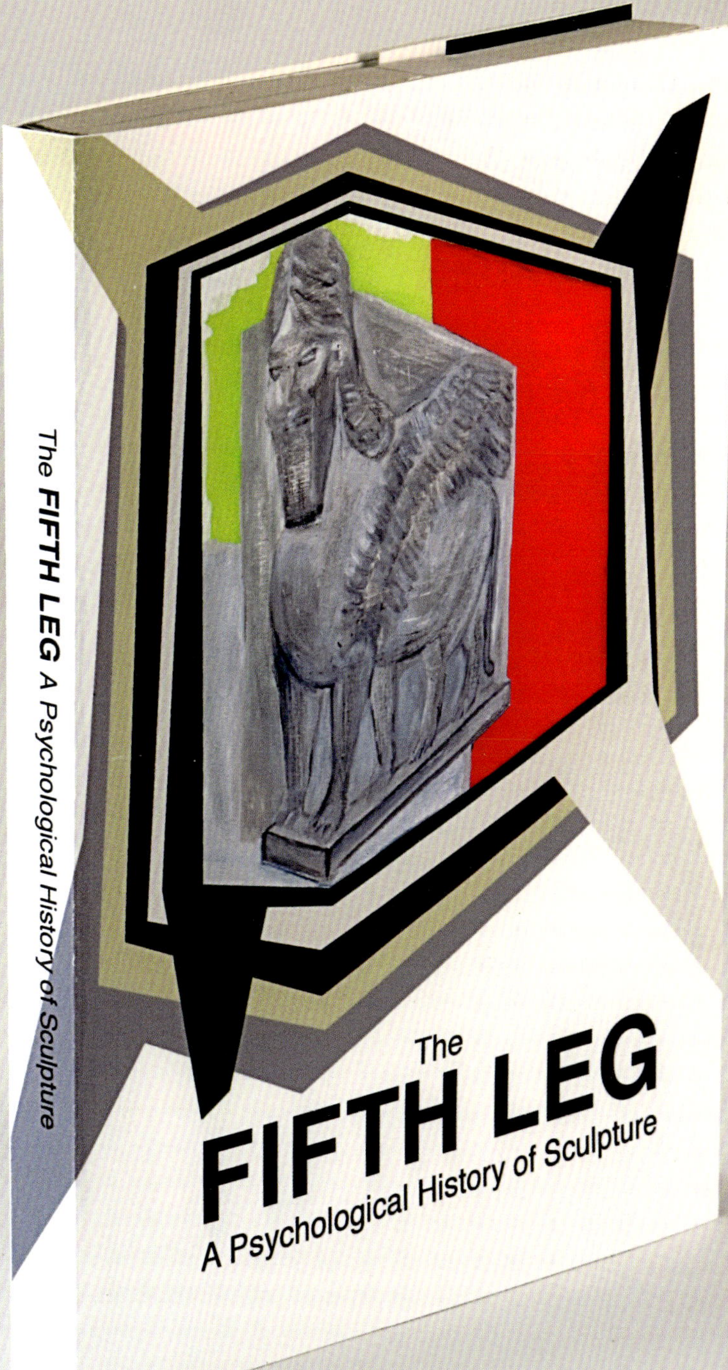
The FIFTH LEG A Psychological History of Sculpture
The
FIFTH LEG
A Psychological History of Sculpture

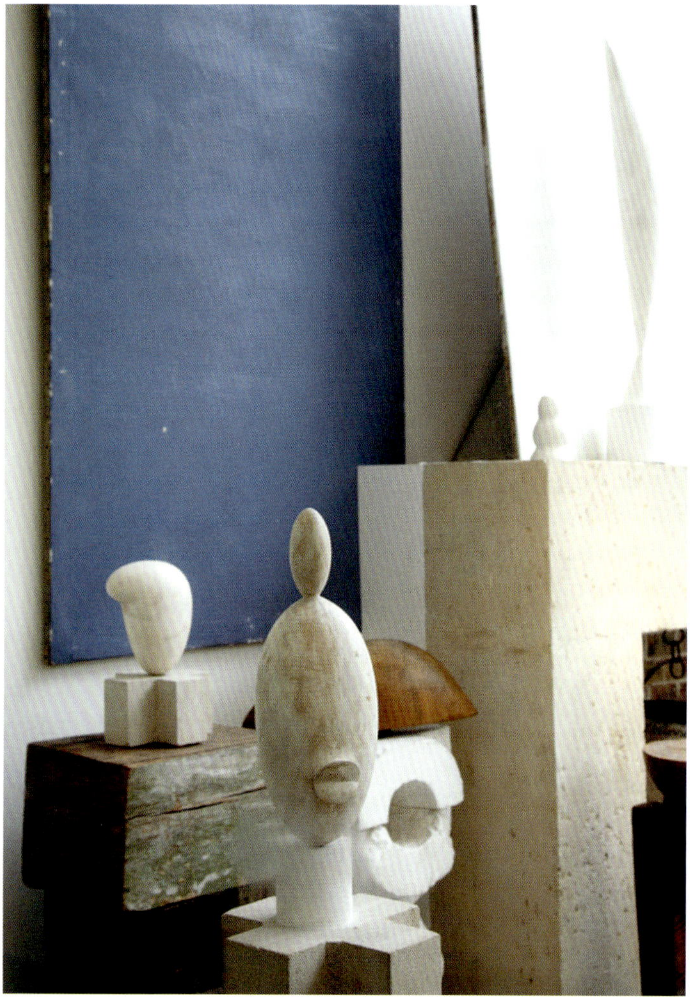

Brancusi's studio, Paris
Photograph by
Leslie Williamson

(facing)

Installation view, *Problem Picture Sources New Sculptures*, Rachel Uffner Gallery, New York, 2010

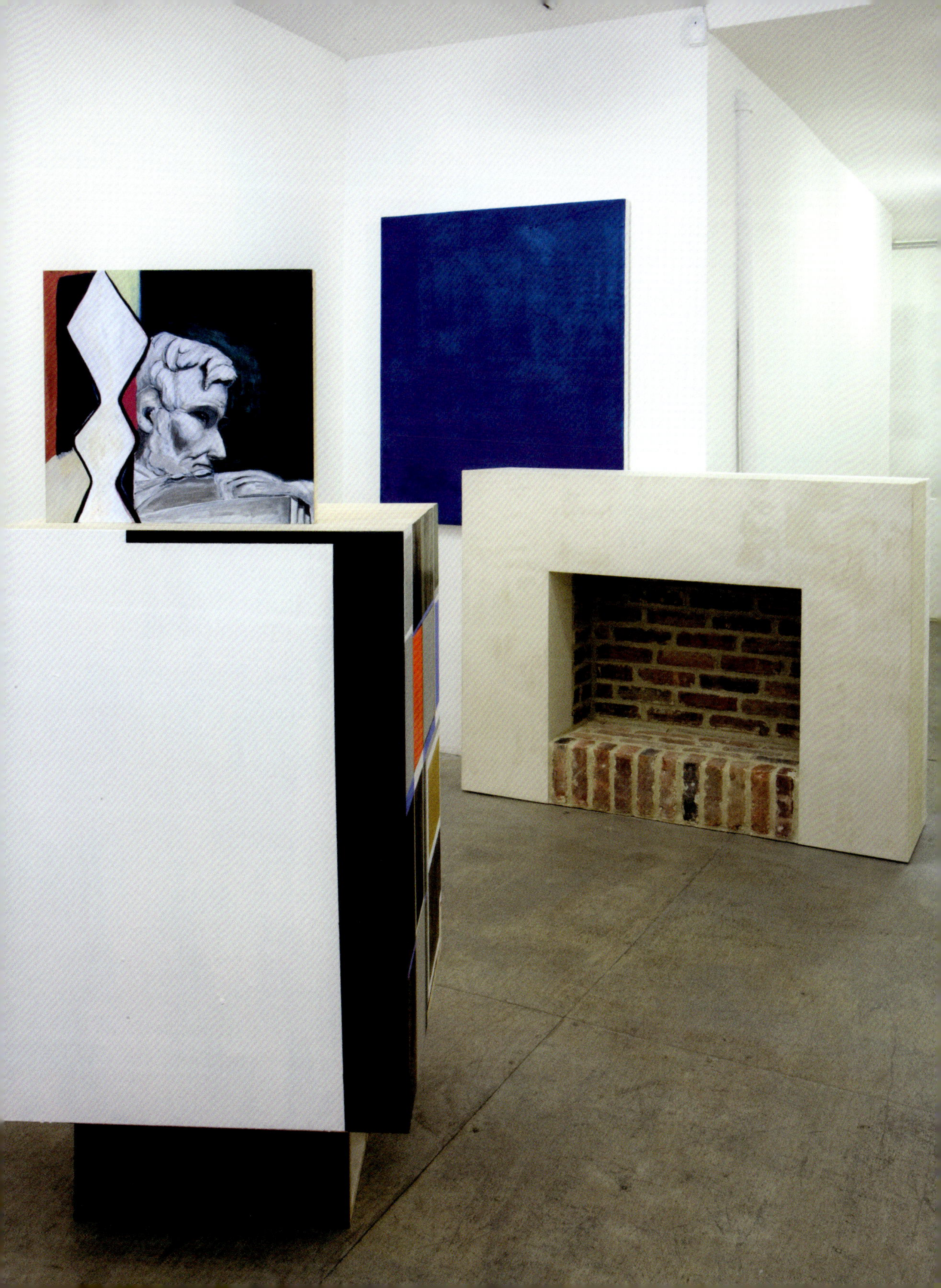

Lincoln bookend obstruction,
2010
Acrylic, paint, panel, ACX
plywood, plaster
60 x 23½ x 23½ inches

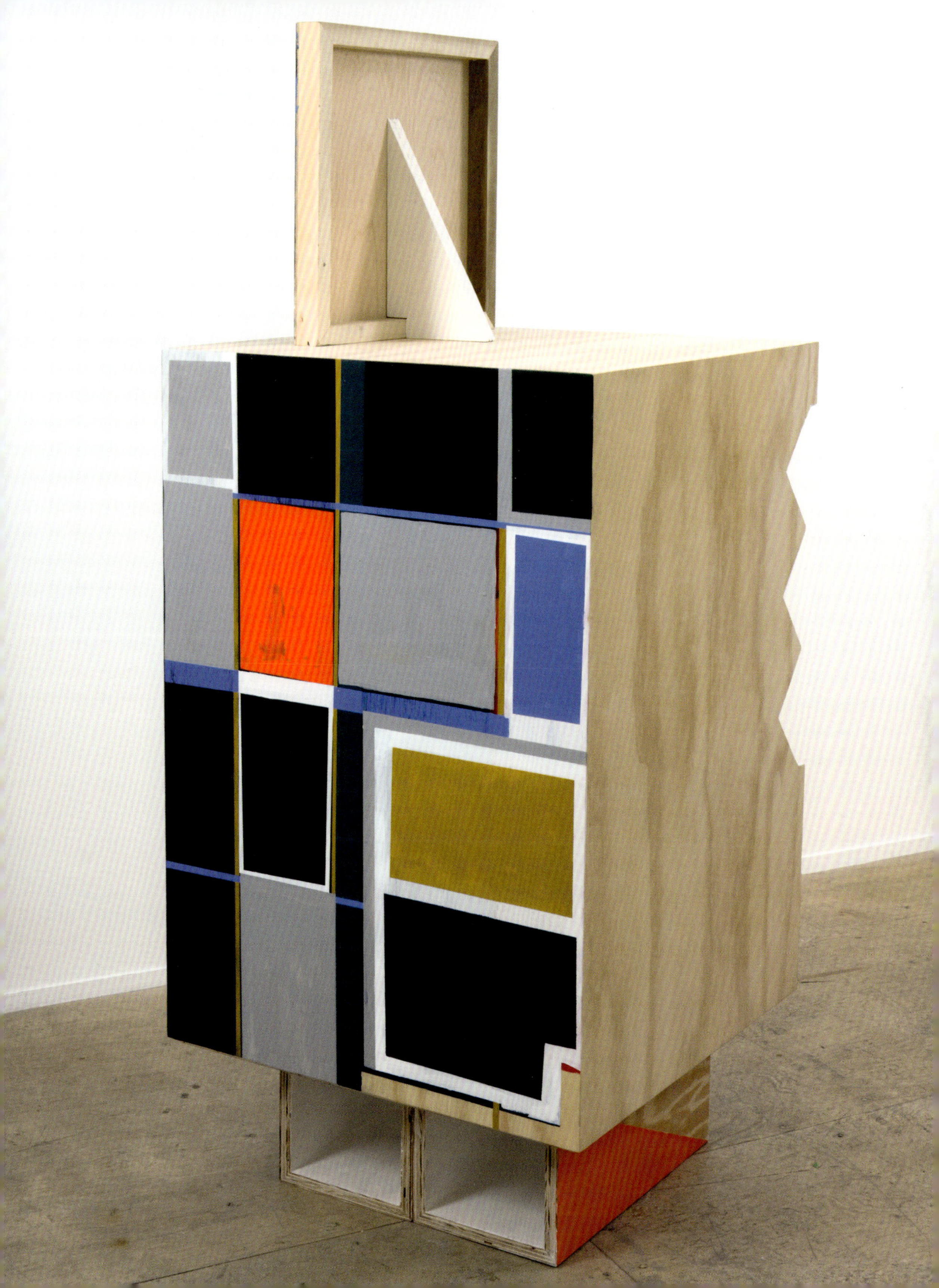

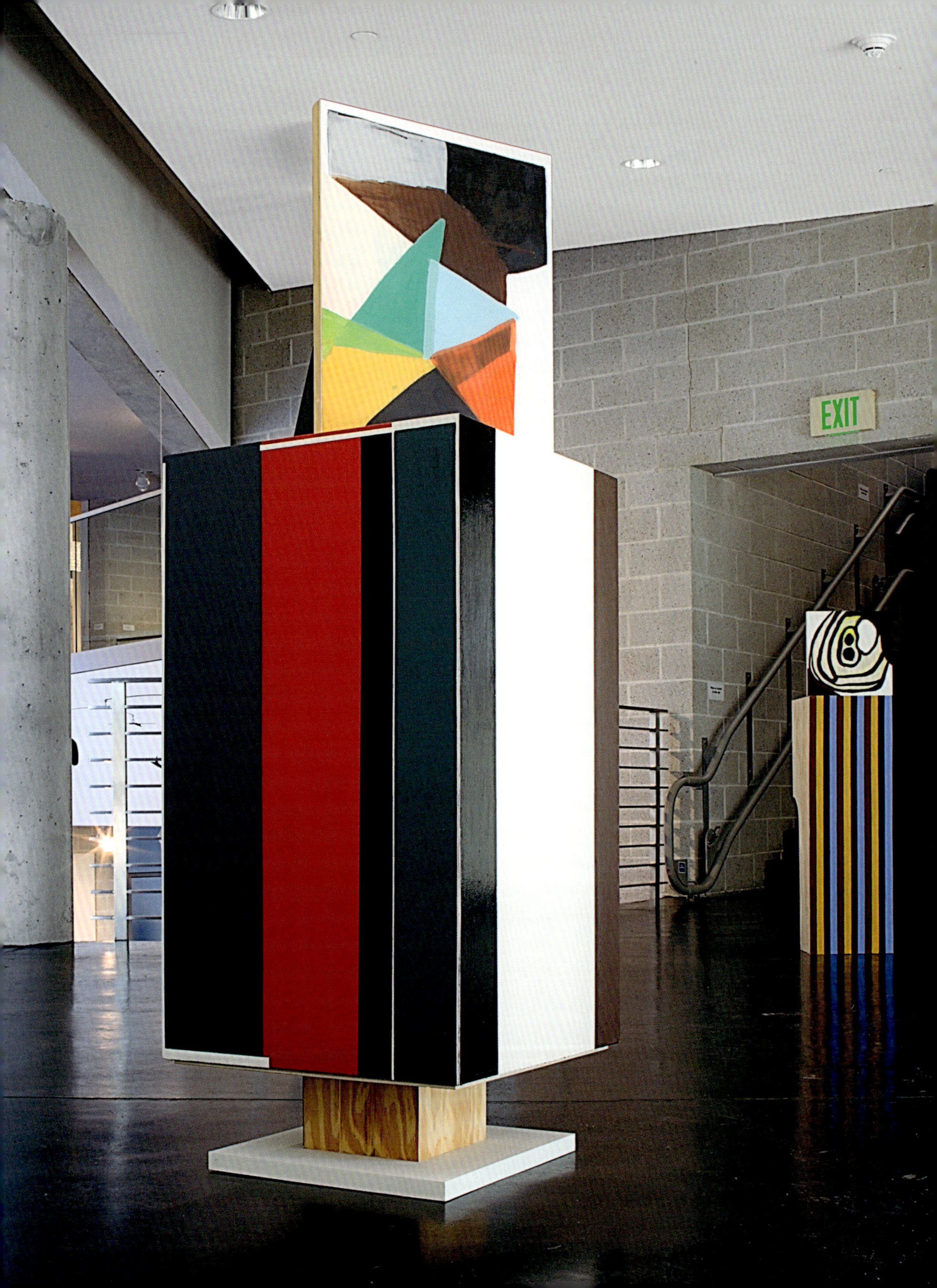
EXIT

CHECKLIST

All works by
Pam Lins

Dimensions listed in inches
height × width x depth

All works courtesy
of the artist
and Rachel Uffner
Gallery, New York

1
Corner piece reconstructed painting background, 2005/2012
Plywood, paint, paintings
Installation dimensions variable, approx 4 × 8 feet

2
Folks straggling along vast stretches of concrete, heading into the plane., 2008
Acrylic, panel, plywood, plaster
63½ × 15¾ × 25½ inches

3
The leaves are falling. Point things out., 2008
Acrylic, panel, plywood, plaster
48 × 11½ × 11½ inches

4
Fireplace reconstructed sculpture background, 2010
Painting (transcribed by Jessica Dickenson), bricks, mortar, plaster, medium-density fiberboard
73 × 73 × 17 inches

5
Medusa page design decoy, 2010
Spray paint, acrylic, paint, panel, ABX plywood, plaster
66 × 20 × 20 inches

6
Statue painting delays, 2010
Acrylic, paint, panel, ABX plywood, plaster
63 × 19½ × 20 inches

7
Slabs and Armatures, Ascot Silver, 2012
Plaster, papier-mâché, pencils, waxed cord, aluminum wire, auto lacquer, latex primer, medium-density fiberboard, ABX plywood
68 × 24 × 24 inches

8
Slabs and Armatures, Denver Gold, 2012
Plaster, papier-mâché, pencils, waxed cord, aluminum wire, auto lacquer, medium-density fiberboard, ABX plywood
55 × 22 × 22 inches

9
Slabs and Armatures, Starlight Black, 2012
Plaster, papier-mâché, pencils, waxed cord, aluminum wire, auto lacquer, medium-density fiberboard, ABX plywood
60 × 26 × 16 inches

Installation view, Tang Museum, Saratoga Springs, New York, 2012

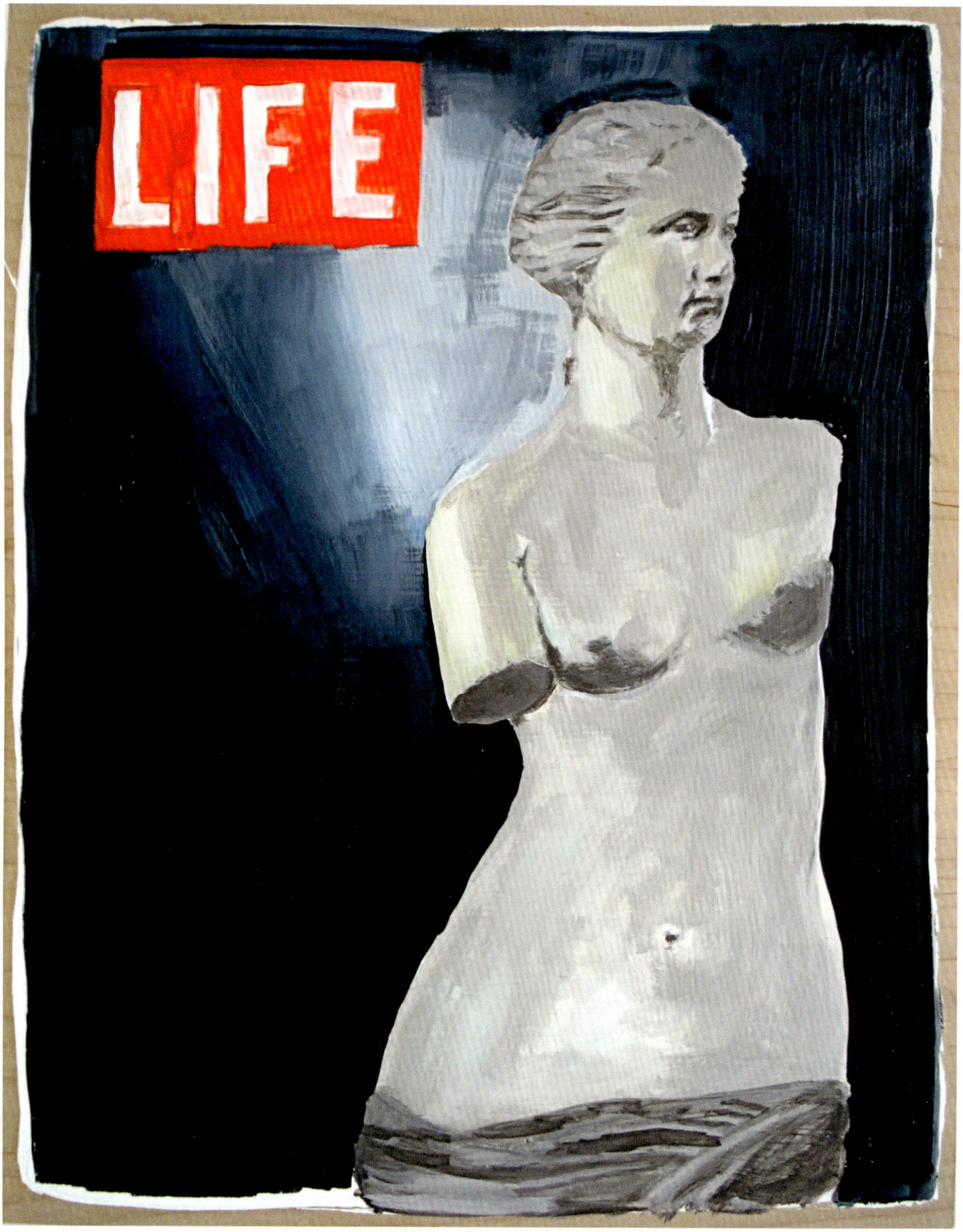
LIFE

PAM LINS

BORN IN CHICAGO, ILLINOIS IN 1957
LIVES AND WORKS IN BROOKLYN, NEW YORK

EDUCATION

1995
M.F.A., Hunter College, City University of New York, New York

1983
B.A., Bemidji State University, Bemidji, Minnesota

SOLO EXHIBITIONS

2012
Opener 23: Pam Lins: Denver Gold, The Frances Young Tang Teaching Museum and Art Gallery at Skidmore College, Saratoga Springs, New York, February 18–April 22

Minus Space, The Suburban, Oak Park, Illinois, January 22–February 27

2010
Problem Picture Sources New Sculptures, Rachel Uffner Gallery, New York, September 12–October 24

2009
Owl, Rachel Uffner Gallery, New York, January 4–February 8

2005
Win Me Free, Silo, New York, October 26–December 4

2003
In Spite of This, Ten In One Gallery, New York, April 26–June 7; traveled to: Mercer Union—Centre for Contemporary Art, Toronto, Canada, July 3–August 2

2001
Hot Snow, Ten in One Gallery, New York

1999
Vent View, Ten in One Gallery, New York

GROUP EXHIBITIONS

2012
you, your sun and shadow, Anderson Gallery at Virginia Commonwealth University, Richmond, Virginia, January 20–March 11

2011
Drawing Club, SOLOWAY, Brooklyn, New York, November 6–November 20

Grand Tour Low Cost, Spazio Morris, Milan, Italy, September 30–November 10

The Idea of the Thing That it Isn't, Halsey McKay Gallery, East Hampton, New York, August 5–September 4

Pop-Up Gallery Portraits, Scot Cohen Realty, Hudson, New York, July 30–September 3

Forcemeat, Wallspace, New York, June 28–August 5

2010
Stuff Art, U·turn Art Space, Cincinnati, Ohio, November 6–27

Parts & Labor, SOLOWAY, Brooklyn, New York, August 15–September 19

2009
NADA's County Affair, Brooklyn, New York, October 18–January 9, 2010

On From Here, Guild & Greyshkul, New York, February 5–8

2008
Without Walls, Museum 52, New York, December 12–January 17, 2009

The Grey House That Thinks Itself Into Your Head Without Asking, Peter Fingesten Gallery at Pace University, New York, September 4–October 2

We, Lizabeth Oliveria Gallery, Los Angeles, California, July 5–August 16

2007
Jerks, Balks, Outblurts, and Jump-Overs, Raid Projects, Los Angeles, California, December 1–January 17, 2008

Public Park, Scoops, Los Angeles, California, October 6–November 3

NeoIntegrity, Derek Eller Gallery, New York, July 19–August 24

Life, 2009
Acrylic on panel
14 x 11 inches

Ceci n'est pas...(This is not...), Sarah Meltzer Gallery, New York, June 27–August 17

Corridor at Sea: The Performative Object, Institute for Contemporary Art at Maine College of Art, Portland, Maine, June 15–August 12

The Line of Time, and the Plane of Now, Harris Lieberman and Wallspace, New York, May 31–June 30

2005
Hunch & Flail, Artists Space, New York, June 21–July 29

2004
A.C., Elizabeth Dee Gallery, New York, July 28–September 2

Field: Science, Technology and Nature, Socrates Sculpture Park, Queens, New York, May 16–August 1

Great Adventure, Roger Bjorkholmen Gallery, Stockholm, Sweden, February 5–March 6

2002
Recreation/Re-Creation: Fun with Found Objects, Noyes Museum of Art, Oceanville, New Jersey, January 12–April 28

2001
Wattage and Friendship, Müller Dechiara, Berlin, Germany, October 4–November 17

2000
New York Projects, Delfina Arts, London, England, July 27–September 10

The Thrill of it All, Ten in One Gallery, New York

Achieving Failure: Gym Culture 2000, Thread Waxing Space, New York, March 23–April 29; traveled to: Cincinnati Art Museum, Cincinnati, Ohio

1999
Pam Lins and Carol Jackson, Ten in One Gallery, New York

Hang Time, White Columns, New York, June 25–August 1

Paradise 8, Exit Art, New York, January 16–April 3

Answer Yes, No, or I Don't Know, Andrew Kreps Gallery, New York, January 6–February 27

1998
Pamela Lins & Liza Phillips, Momenta Art, Brooklyn, New York, November 1–30

Hey, You Never Know, 534 LaGuardia Place, New York, March 3–April 30

1997
Sarah Staton's Supastore, Cornerhouse, Manchester, England, October 4–November 16

Working In Brooklyn: Current Undercurrent, Brooklyn Museum of Art, Brooklyn, New York, July 25–January 25, 1998

1995
Split Level, Art in General, New York, March 18–April 29

1991
Object Lesson, BRIC Rotunda Gallery, Brooklyn, New York, September 26–November 2

1990
Suggestive Objects, Art in General, New York, April 7–May 5

BIBLIOGRAPHY

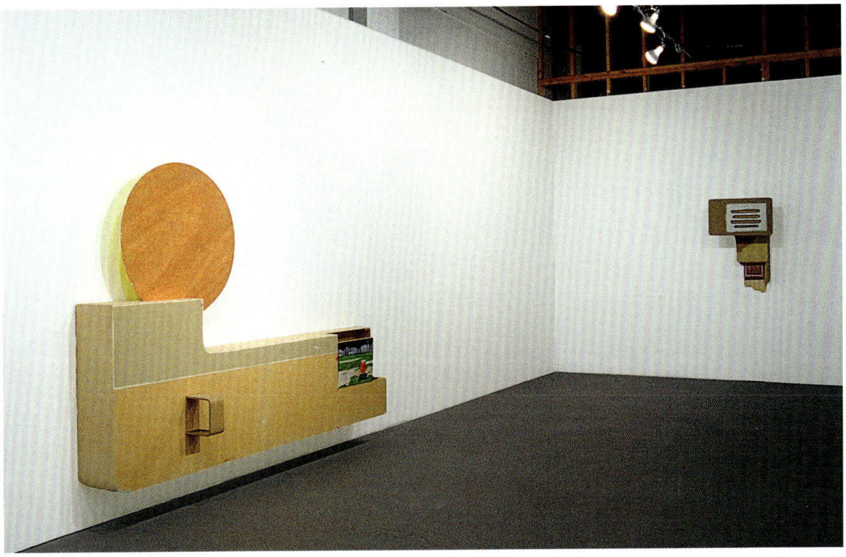

Installation view, *In Spite of This*, Mercer Union—Centre for Contemporary Art, Toronto, 2003

SELECTED BOOKS AND CATALOGUES

Arning, Bill, and Joel Sanders. *Achieving Failure: Gym Culture*. Exhibition catalogue. New York: Thread Waxing Space, 2000.

Berry, Ian. *Opener 23: Pam Lins: Denver Gold*. Exhibition catalogue. Saratoga Springs, New York: The Frances Young Tang Teaching Museum and Art Gallery at Skidmore College, 2012.

Burton, Johanna, and Steve Lefreniere. *Pam Lins: In Spite of This*. Exhibition catalogue. New York: Ten in One Gallery, 2003.

Fry, Naomi. "The Time and Temperature Sculpture." In *Corridor at Sea: The Performative Object*. Exhibition catalogue. Portland, Maine: Institute of Contemporary Art at Maine College of Art, 2007: 9–16.

Hang Time: Christian Alexa, Janet Cohen, Neil Goldberg, Pam Lins, Sylvan Lionni, Eric Porter, George Raggett, Andrea Ray, Type A: White Columns, June 25–August 1, 1999. New York: White Columns Gallery, 1999.

I Like Your Work: Art and Etiquette. Brooklyn, New York: Paper Monument, 2009.

Lyall, Scott, and Janis Demkiw. *Pam Lins, Derek Sullivan, Adam Frelin*. Exhibition brochure. Toronto, Canada: Mercer Union Gallery, 2003.

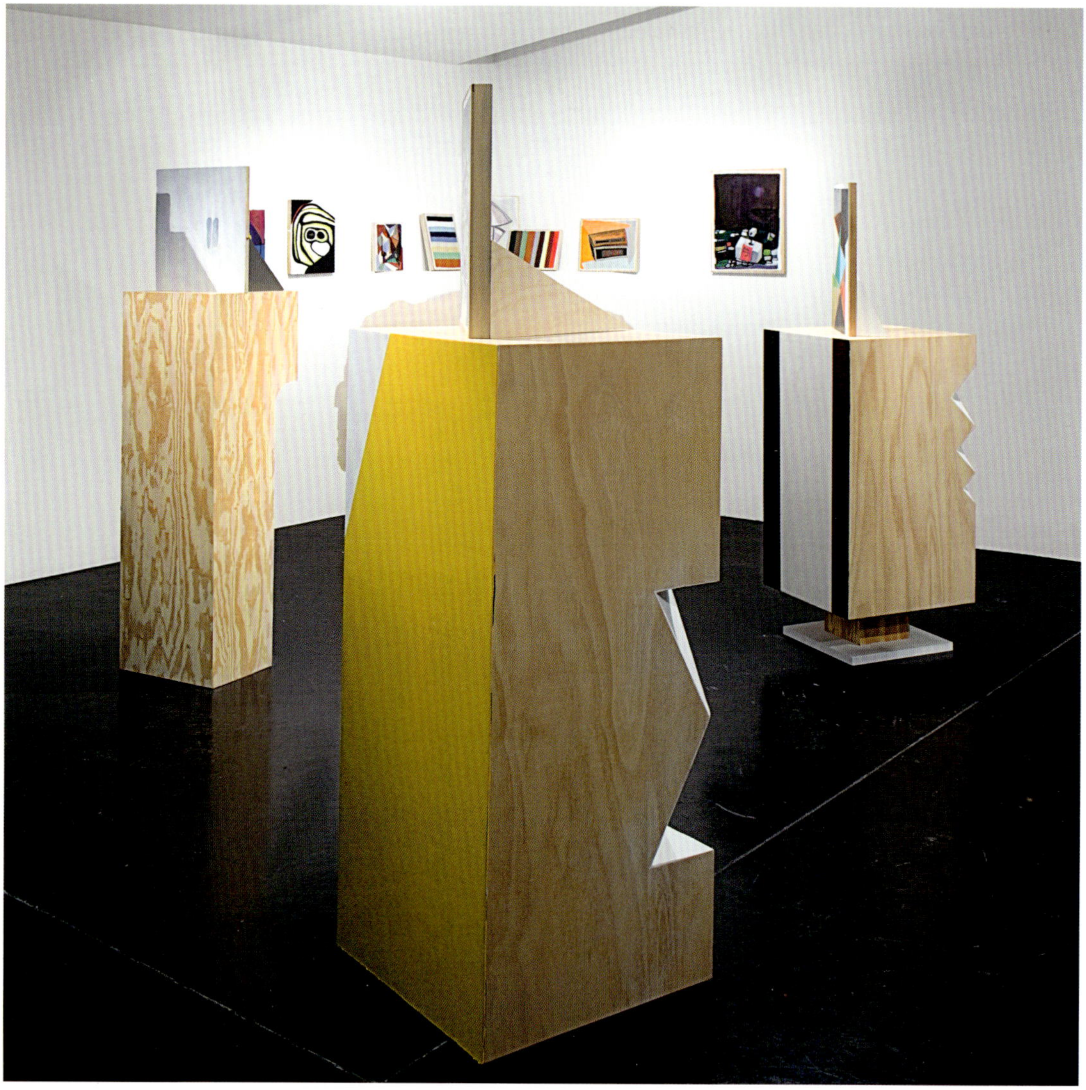

Installation view, Tang Museum, Saratoga Springs, New York, 2012

SELECTED ARTICLES AND REVIEWS

Adler, Dan. "Pam Lins: Ten in One Gallery, New York." *Zing Magazine* 17 (September 2002).

Arning, Bill. "'Achieving Failure': Gym Culture 2000." *The Gay & Lesbian Review Worldwide* 7, no. 1 (31 January 2000): 39.

——. "Hang Time." *Time Out New York* (29 July–5 August 1999): 51.

Banai, Nuit. "Pam Lins." *Time Out New York* (29 January–4 February 2009): 50.

Boucher, Brian. "Pam Lins: Rachel Uffner." *Art in America* 98, no.11 (December 2010): 148–149.

Brody, David. "The Fifth Leg: Pam Lins at Rachel Uffner Gallery." *artcritical* (22 October 2010) http://www.artcritical.com/2010/10/22/pam-lis/ (accessed 4 April 2012).

Burton, Johanna. "Not a Single Point of View: Contemporary Sculpture and the Spatial Imaginary." *Yale University Art Gallery Bulletin* (2009): 70–83.

Coburn, Tyler. "Pam Lins / Owl." *ArtReview* 31 (April 2009): 97.

Cotter, Holland. "Go Ahead, Expect Surprises." *New York Times* (9 August 2007): E1.

Feinstein, Rochelle. "Artists on Artists: Rochelle Feinstein on Pam Lins." *BOMB Magazine* 100 (Summer 2007): 22–23.

"Front Gallery: Pam Lins, Back Gallery: Derek Sullivan." *Mercer Union: A Center for Contemporary Art* (3 July–2 August 2003).

"Hunch and Flail." *New Yorker* (1 August 2005): 16.

Hunt, David. "Pumping Irony." *Time Out New York* (13–20 April 2000).

Johnson, Ken. "Field-Science, Technology and Nature: Review." *New York Times* (23 July 2004): E31.

——. "Art Review: Hey, You Never Know." *New York Times* (17 April 1998): E39.

Lins, Pamela. "Artists on Artists: Pamela Lins on Andrea Blum." *BOMB Magazine* 117 (Fall 2011): 74–75.

McClister, Nell. "Pam Lins: Ten in One Gallery." *Artforum* 42, no. 2 (October 2003): 173.

Moffatt, Laura. "New York Project: Delfina Project Space, London." *Art Monthly* 239 (September 2000): 48–49.

"Pam Lins." *The New Yorker* (9 February 2009): 20.

Penn, Beverly. "Narrative: The Impulse in Contemporary America Metal." *Metalsmith* 13 (1993): 16-29.

Reden, Lisa. "Pamela Lins: A Kept Object." *Metalsmith* 10 (Spring 1990): 44.

Roberts, Alison. "A Brooklyn Cheer for the British Art Scene." *London Evening Standard* (1 August 2000).

Rosenberg, Karen. "Pam Lins: Problem Picture Sources; New Sculptures." *New York Times* (8 October 2010): C30.

Schmerler, Sara. "Answer Yes, No, or Don't Know. *Time Out New York* (18–25 February 1999).

——. "Hey, You Never Know." *Time Out New York* (23–30 April 1998).

Schwendener, Martha. "Picks: Pam Lins." *Artforum online* (2 December 2005).

Smith, Roberta. "Art Review: Brooklyn Haven for Art Heats Up." *New York Times* (6 November 1998): E33.

Wilson, Michael. "Pam Lins: Rachel Uffner Gallery." *Artforum* 47, no. 7 (March 2009): 242–243.

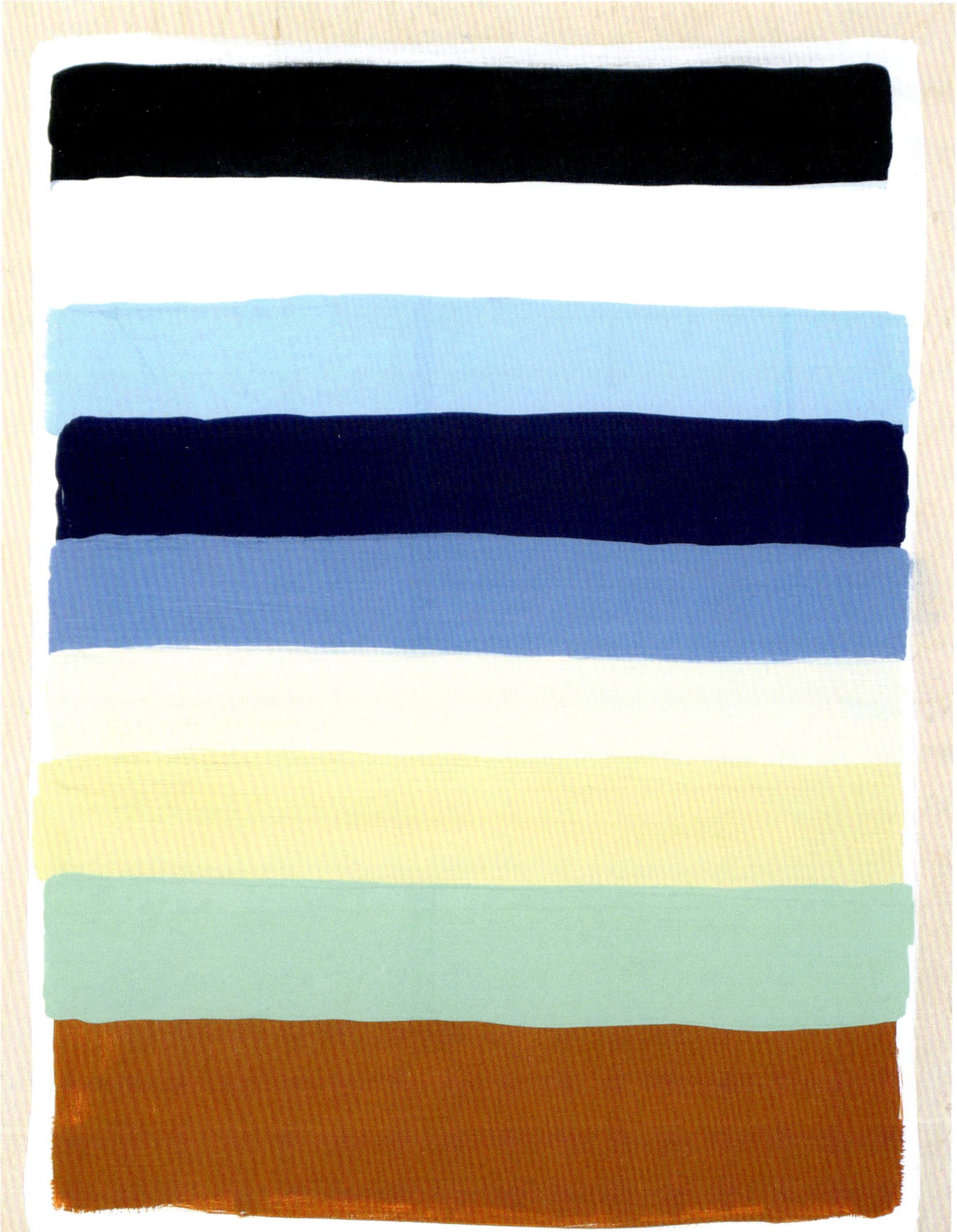

ACKNOWLEDGMENTS

CRAFTING THIS INSTALLATION by Pam Lins provided a puzzle in how best to invite viewers to walk around a space using only objects. A seemingly simple task was made complex as we worked together in the gallery auditioning sculptures in various combinations. All permutations were engaging in different ways, and the exercise put the power of her artwork into high relief. What seemed straightforward about her sculptures and paintings wasn't, and conversely what seemed resistant turned out to be consistently generous.

Thanks to the New York State Council on the Arts, the Overbrook Foundation and the Friends of the Tang for their ongoing support of the Opener series that made this project possible. Special thanks to Rachel Uffner and Amber Vilas of Rachel Uffner Gallery for their support and to Halsey Rodman for his important feedback and encouragement.

Thanks to the many photographers whose images are reproduced in this book and to designer Beverly Joel and editors Jay Rogoff and Megan Hyde. The entire Tang Museum staff deserves appreciation: thanks to Ginger Ertz, Torrance Fish, Megan Hyde, Elizabeth Karp, Susi Kerr, Chris Kobuskie, Ryan Lynch, Jennifer Napierski, Patrick O'Rourke, Lorri Riggs, Vickie Riley, Barbara Schrade, Rachel Seligman, Kelly Ward, John Weber, and our installation crew, Luke Anderson, Samuel Coe, Patrick Girard, Scott Heydrick, Matthew Scellen, and Cynthia Zellner. Thanks also to our student interns Naomi Brown '12 and Alexander Unkovic '12.

Pam Lins was a great collaborator—offering ideas, answers, and challenges at every turn. Thanks to her for the loan of artwork, including several new works on view for the first time, and her critical assistance as we re-created older works. The finished exhibition was itself a new work and marks an important moment as we realized the experimental, laboratory-like potential of our museum. —Ian Berry

THANKS TO Ian Berry and everyone at the Tang, Rachel Uffner, and Halsey Rodman. —Pam Lins

1974 Delay, 2012
Acrylic on panel
12 x 9 inches

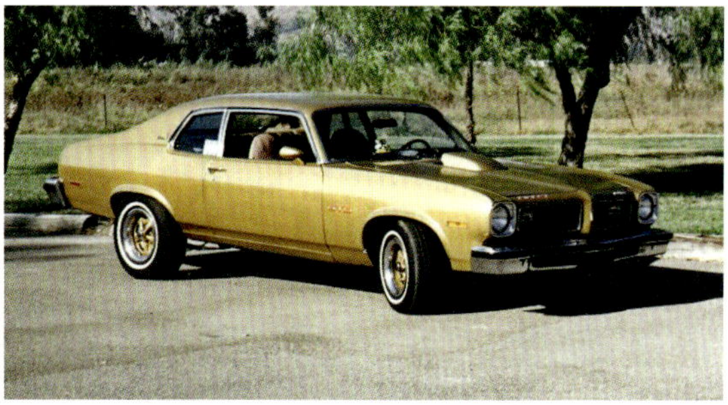

This catalogue accompanies the exhibition

OPENER 23
PAM LINS: DENVER GOLD

The Frances Young Tang Teaching Museum and Art Gallery at Skidmore College
Saratoga Springs, New York
February 18–April 22, 2012

The Frances Young Tang Teaching Museum and Art Gallery
Skidmore College
815 North Broadway
Saratoga Springs, New York 12866
T 518 580 8080
F 518 580 5069
www.skidmore.edu/tang

State of the Arts
NYSCA

This exhibition and publication are made possible in part with public funds from the New York State Council on the Arts, a state agency, The Overbrook Foundation, and the Friends of the Tang.

All Rights Reserved. No part of this publication may be reproduced, stored, or transmitted in any form or by any means without permission in writing from the Tang Teaching Museum.

© 2013 The Frances Young Tang Teaching Museum and Art Gallery
ISBN: 9780982148662
Library of Congress Control Number: 2013949267

All images in this catalogue, unless otherwise noted, are copyright of the artist and courtesy of the artist and Rachel Uffner Gallery. Thanks to all the photographers whose work is reproduced in the book including those listed below:

Pages 1, 14, 18, 19, 20, 34, 36–37: Pam Lins
Pages 2, 4, 23, 39, 40–41, 43, 48, 51: Adam Reich
Pages 9, 44, 45, 56, 62: Courtesy of Rachel Uffner Gallery
Pages 25, 26, 27: John Berens
Pages 28–29: Courtesy of Ten in One Gallery
Pages 32–33, 54, 60: Arthur Evans
Pages 46–47, 49: Cliff Boress
Page 50: Leslie Williamson
Page 59: Brian Piitz, Courtesy of Mercer Union

Designed by Beverly Joel, pulp, ink.

Printed in Italy by Conti Tipocolor

Cover:
Lincoln bookend obstruction, 2010 (detail)
Acrylic, paint, panel, ACX plywood, plaster
60 x 23½ x 23½ inches

Page 1:
Photograph by Pam Lins

Page 2 and 4:
Studio views, 2012

left to right:
Slabs and Armatures, Denver Gold, 2012
Plaster, papier-mâché, pencils, waxed cord, aluminum wire, auto lacquer, medium-density fiberboard, ABX plywood
55 x 22 x 22 inches

Slabs and Armatures, Ascot Silver, 2012
Plaster, papier-mâché, pencils, waxed cord, aluminum wire, auto lacquer, latex primer, medium-density fiberboard, ABX plywood
68 x 24 x 24 inches

Slabs and Armatures, Starlight Black, 2012
Plaster, papier-mâché, pencils, waxed cord, aluminum wire, auto lacquer, medium-density fiberboard, ABX plywood
60 x 26 x 16 inches